The Physical and Chemical Basis of

Molecular Biology:

Fundamentals

The Physical and Chemical Basis of
Molecular Biology:
Fundamentals

Thomas E. Creighton

HP

Helvetian Press

Published by Helvetian Press 2011

www.HelvetianPress.com

HelvetianPress@gmail.com

ISBN: 978-0-9564781-3-9

CONTENTS

The current volume consists of the first six chapters of

The Physical and Chemical Basis of Molecular Biology

PREFACE

The field of molecular biology continues to be the most exciting and dynamic area of science and is predicted to dominate the 21st century. Only by investigating biological phenomena at the molecular level is it possible to understand them in detail. Such understanding is vital for advances in medicine, and the pharmaceutical industry that produces new drugs and cures is greatly dependent upon molecular biology. Molecular biology also contributes to our understanding of what human beings are and how they fit into this universe. Comparing the amino acid and nucleotide sequences of humans with those of other organisms can only confirm that humans are one very small part of the living world.

Proteins and nucleic acids are the primary subjects of molecular biology. They carry, transmit, and express the genetic information that defines each living organism. At the heart of molecular biology are the techniques that are used to understand these complex macromolecules. Nucleic acids have the great advantage that the members of each type behave virtually identically, irrespective of their nucleotide sequence but dependent primarily on their length. Consequently, the same techniques are likely to succeed with any of them, and recipes and kits are available for many of the routine experiments and measurements. In contrast, proteins are highly individualistic, and many techniques usually need to be varied to be applied to any specific protein. In this case, it is vital to understand the physical and chemical basis of the techniques. Even in the case of nucleic acids, one should be aware of how and why the technique works, and when it does not, so as not to blunder into mistaken interpretations of results obtained by simply following a recipe.

This volume attempts to provide the background of which every molecular biologist should be aware. It is the book that I wish had been available throughout my career.

The first six chapters describe briefly some of the more fundamental aspects. Thermodynamics is central to understanding the stabilities and energetics of macromolecules and the reactions and interactions that they undergo; only those aspects of this immense subject that are pertinent to molecular biology are presented here (Chapter 1). Molecular biology is not concerned with macromolecules in isolation, but with their interactions with other molecules (Chapter 2). The physical aspects of these interactions in isolation are understood in detail, but those in molecular biology generally occur within cells, within an aqueous environment, and the amazing properties of water are involved in all of them. It is vital to understand aqueous solutions (Chapter 3). Life is a dynamic phenomenon, so the rates at which reactions occur is of crucial importance (Chapter 4). So many techniques in molecular biology use radioactivity that one should be aware of its fundamental properties (Chapter 5). The sizes of macromolecules vary enormously and are their hallmarks. The most accurate and powerful method of measuring the sizes of molecules is mass spectrometry, which can often identify molecules simply on that basis, and it has become central to all studies of proteins and nucleic acids (Chapter 6).

The next three chapters deal with how to visualize the structures of macromolecules using their interactions with light of widely varying wavelengths. Macromolecules in solution scatter radiation and thereby reveal information about their structures (Chapter 7). Immobilized macromolecules can be observed directly in microscopes, using either visible light or electrons; their physical surfaces can also be sensed using very sensitive probes (Chapter 8). When arranged in a crystalline array, their most intimate structural details can be visualized from how they scatter and diffract X-rays or neutrons (Chapter 9).

Spectroscopic techniques that monitor the interactions of radiation with molecules are amongst the most useful in molecular biology and are described in the following five chapters. Most commonly used are the absorption and emission of visible and UV light (Chapter 10). The interaction of polarized light with molecules depends critically upon their chiral properties and is a very useful probe of molecular structure (Chapter 11). The absorption of light by stimulation of the vibrational properties of molecules can be very informative about their structures (Chapter 12). Nuclear magnetic resonance (NMR) complements X-ray diffraction, in that it also reveals the detailed structures of macromolecules, but while dissolved in aqueous solution, and also provides unique information about their dynamic properties (Chapter 13). Some of the most important biological reactions are involved in the transfer of electrons from one molecule to another, and this often produces free radicals with unpaired electrons that give the molecules electron magnetic resonance properties (Chapter 14).

Most other techniques of molecular biology involve the transport of macromolecules in solution and are described in the following four chapters. The rates at which macromolecules move in solution are determined by their sizes and shapes (Chapter 15). Molecules can be induced to sediment by applying a centrifugal force, and the rates at which they do so also provide information about their sizes and shapes (Chapter 16). Proteins and nucleic acids usually have overall net electrical charges, due to ionized groups, so they can be induced to migrate in an electrical field; such electrophoretic techniques are central to molecular biology (Chapter 17). The large sizes of macromolecules can make it impossible for them to enter pores of molecular sieves, which can provide information about their sizes and also permit their separation from molecules of other sizes (Chapter 18).

The last three chapters describe the most fundamental functional properties of proteins and nucleic acids: their interactions with other molecules (Chapter 19). The interactions of macromolecules with solid supports provide a great variety of methods of separating them using chromatographic techniques (Chapter 20). The large sizes of these macromolecules make it possible to retain at least some of their functional properties while adsorbed to a solid support, and techniques that make use of this are some of the most important in molecular biology (Chapter 21).

The 21 chapters in this volume provide a comprehensive description of the chemical and physical basis of most of molecular biology. Of course, not all techniques could be described fully, and it is unfortunate that some of the most sophisticated techniques, which would require the greatest discussion, are not the most important to the average molecular biologist and have not been treated in detail. I have tried to match the degree of description given to the importance of the subject to the average molecular biologist. Otherwise, this work would have been much longer and impractical.

The references listed were chosen to be those that would best provide the interested reader with entry to the literature. They should not be assumed to be those most important for the subject.

No one person can be expert in all the techniques of molecular biologist, but I have had the good fortune to write two editions of a comprehensive book on proteins, *Proteins: Structures and Molecular Properties*, published by W.H. Freeman. I have also edited two editions of *Protein Structure: a practical approach* and of *Protein Function: a practical approach*, published by IRL Press, and two multi-volume encyclopedias, *Encyclopedia of Molecular Biology* and *Encyclopedia of Molecular Medicine*, published by Wiley-Interscience. The information available in these volumes has been invaluable while preparing the present volume. I have made ample use of the work of others more expert than me but too numerous to list. Of course, shortcomings and errors in this volume are totally my responsibility, for which I apologize in advance. Corrections, criticisms and suggestions would be welcome and can be sent to me at HelvetianPress@gmail.com. Special thanks are due to R. W. Woody for pointing out errors in the original manuscript.

Hopefully, the references to proteins and nucleic acids throughout this volume will have whetted your appetite to learn more about these fascinating macromolecules. Much more information is available in the companion volume, *The Biophysical Chemistry of Nucleic Acids and Proteins.*

Thomas E. Creighton

COMMON ABBREVIATIONS

a:	atto (10^{-18})
Å:	Ångstrom (= 0.1 nm)
A:	adenine
ac:	alternating current
ADP:	adenosine diphosphate
Ala:	alanine residue of a protein
AMP:	adenosine monophosphate
Arg:	arginine residue of a protein
Asn:	asparagine residue of a protein
Asp:	aspartic acid residue of a protein
ATP:	adenosine triphosphate
BPTI:	bovine pancreatic trypsin inhbitor
C_p:	heat capacity at constant pressure
C:	cytidine
cal:	calorie (= 4.184 joules)
CD:	circular dichroism
cDNA:	complementary DNA
CDP:	cytidine diphosphate
cmc:	critical micelle concentration
CMP:	cytidine monophosphate
CTP:	cytidine triphosphate
Cys:	cysteine residue of a protein
Da:	Dalton
dc:	direct current
ddNTP:	dideoxynucleoside triphosphate
DNA:	deoxyribonucleic acid
dNTP:	deoxynucleoside triphosphate
e:	mathematical number (2.718) that is the base of natural logarithms
e:	unit of atomic charge (1.602×10^{-19} C)

EDTA: ethylenediamine-*N,N,N',N'*-tetracetic acid
EGTA: ethyleneglycol bis(β-aminoethyl ether) *N,N,N',N'*-tetracetic acid
EPR: electron paramagnetic resonance
ESR: electron spin resonance
Et: ethyl group ($-CH_2 - CH_3$)
f: femto (10^{-15})
FPLC: fast-protein liquid chromatography
g: gram
g: gravitational constant (9.81 m s^{-2} or 6.673 x 10^{-11} N m^2/kg^2)
G: Gibbs free energy
G: Guanine
G: Giga (10^9)
Gdm: guanidinium
GDP: guanosine diphosphate
Gln: glutamine residue of a protein
Glu: glutamic acid residue of a protein
Gly: glycine residue of a protein
GMP: guanosine monophosphate
GSH: glutathione, thiol form
GSSG: disulfide form of glutathione
GTP: guanosine triphosphate
H: enthalpy
h: Plank's constant (1.584 x 10^{-34} cal s; 6.626 x 10^{-34} J s)
His: histidine residue of a protein
HPLC: high-performance liquid chromatography
Ile: isoleucine residue of a protein
IPG: immobilized pH gradient
IR: infrared
J: joule
K: absolute temperature
k_B: Boltzmann's constant (3.298 x 10^{-24}cal K^{-1} ; 1.381 x 10^{-23} J K^{-1})
k: rate constant
k: kilo (10^3)
K_{eq}: equilibrium constant
K_M: Michaelis constant
l: liter (10^{-3} m^3)
Leu: leucine residue of a protein
Lys: lysine residue of a protein
μ: micro (10^{-6})

m: milli (10^{-3})
m: meter
M: molar (moles/liter)
M: mega (10^{6})
Me: methyl group ($-CH_3$)
Met: methionine residue of a protein
mRNA: messenger RNA
n: nano (10^{-9})
N: Newton (1 kg m s^{-2} = 1 J m^{-3})
N_A: Avogadro's number (6.022 x 10^{23} mol^{-1})
NAD: nicotinamide adenine dinucleotide
NADH: reduced form of NAD
NADP: nicotinamide adenine dinucleotide phosphate
NADPH: reduced form of NADP
n_H: Hill coefficient
NMR: nuclear magnetic resonance
NTP: nucleoside triphosphate
p: pico (10^{-12})
p.p.m.: parts per million
p.s.i.: pounds per sqare inch
PAGE: polyacrylamide gel electrophoresis
PCR: polymerase chain reaction
PDB: Protein Data Bank
Phe: phenylalanine residue of a protein
pI: isoelectric point
P_i: inorganic phosphate
PP_i: inorganic pyrophosphate
Pro: proline residue of a protein
Pu: purine
Py: pyrimidine
r.m.s.: root-mean-square
R: gas constant (1.987 cal mol^{-1} K^{-1} ; 8.315 J mol^{-1} K^{-1}, = $N_A k_B$)
redox: reduction/oxidation
RF: radio-frequency
RNA: ribonucleic acid
r.p.m.: revolutions per minute
S: entropy
S: Svedberg unit of sedimentation (10^{-13} s)
SDS: sodium dodecyl sulfate

Sec.: section
Ser: serine residue of a protein
T: temperature
T: tera (10^{12})
T: Tesla
T: Thymine
TFA: trifluoroacetic acid
Thr: threonine residue of a protein
TLC: thin-layer chromatography
tRNA: transfer RNA
Trp: tryptophan residue of a protein
TTP: thymidine triphosphate
Tyr: tyrosine residue of a protein
U: uracil
UDP: uridine diphosphate
UMP: uridine monophosphate
UTP: uridine triphosphate
UV: ultraviolet
Val: valine residue of a protein
V_{max}: maximum velocity of an enzyme-catalyzed reaction
z: zepto (10^{-21})

~ GLOSSARY ~

Ab initio: starting from the beginning or first principles.

Abscissa: the horizontal axis of a graph.

Achiral: having no chirality, with a plane of symmetry, so that the mirror image is identical.

Acid: any compound that can supply a proton.

Activation energy: the free energy barrier ($\Delta G^{\ddagger}$) that must be overcome for a chemical reaction to occur.

Active site: the region of an enzyme where the substrate binds and the chemical reaction occurs.

Actvity, chemical: concentration corrected for nonideality.

Activity coefficient: constant multiplied by the concentration to give the chemical activity.

Adduct: a chemical group added to another molecule.

Adiabatic: occurring without loss or gain of heat.

Aerobic: in the presence of air or oxygen.

Agonist: substance that produces the same response as a hormone.

Algorithm: set of instructions that define a method, usually a computer program.

Allostery: binding of a ligand at one site on a macromolecule affects another site on the same molecule.

Amphiphile: molecule having both a hydrocarbon part and a polar part, so that it localizes at interfaces between hydrocarbons and water.

Amphoteric: containing both acidic and basic groups.

Anaerobic: in the absence of air or oxygen.

Analyte: molecule being analyzed by some technique.

Anion: negatively-charge molecule.

Anisotropic: exhibiting properties with diff erent values when measured in different directions.

Annealing: association of oligonucleotides by forming base pairs between them.

Anode: positively-charged electrophoresis terminal toward which negatively charged molecules (anions) migrate.

Anomers: the two isomers that result when sugar molecules are linked together, due to the C-1 atom becoming asymmetric, usually designated as alpha or beta.

Antagonist: substance that prevents the response of a hormone.

Antigen: any molecule recognized specifically by an antibody.

Antisense: oligonucleotide complementary in nucleotide sequence to an original strand, designated the "sense" strand.

Apoenzyme: enzyme without its coenzyme.

Aprotic: incapable of donating a proton.

Aptamer: nucleic acid that was selected to bind a specific ligand.

Artifact: something created by humans.

Autoradiography: detection of radioactivity by its effect on photographic emulsions.

Bacteriophage: a virus that multiplies in bacteria.

Base: any compound that can accept a proton.

Bohr effect: influence of pH on the oxygen affinity of hemoglobin.

Boltzmann distribution: the population of a species in equilibrium with others is proportional to the negative exponential of its energy.

Buffer: a mixture of acidic and basic forms of a reagent that tends to keep the pH constant.

Calorie: the amount of heat necessary to raise the temperature of 1 g of water from 15° C to 16° C.

Canonical: conforming to a general rule.

Catalyst: any substance that increases the rate of a chemical reaction without being consumed in that reaction.

Cathode: negatively-charged electrophoresis terminal toward which positive charges migrate.

Cation: positively-charged molecule.

Chaotropic: biologically disruptive.

Chelate: multiple interactions between several groups of a molecule and a metal ion.

Chemical potential: the partial molar Gibbs free energy.

Chemiluminescence: light emission during chemical reactions that results from the decay of excited species.

Chiral: consisting of nonsuperimposable isomers that are mirror images of each other.

Chromatin: genomic DNA and associated proteins, as found in chromosomes.

Chromophore: a molecule or moiety that absorbs light and appears colored.

Chromosome: self-replicating structure of DNA and proteins that contains the genetic information.

Cis-acting: acting on the same molecule.

Clone: replicas of all or part of a macromolecule, or a cell, produced by replication.

Coding region: nucleic acid segment that contains the linear arrangement of codons specifying (by the genetic code) the order of amino acid residues in a protein.

Codon: a three-nucleotide unit in a gene or messenger RNA used by the genetic code.

Coenzyme: a molecule required by a number of different enzymes and used as a substrate, alternating between two forms, such as NAD and NADH.

Cofactor: a molecule required by a number of enzymes for their catalytic activity, but not changed in the reaction.

Cognate: recognized specifically.

Coherent: all the waves have the same phase.

Coherent scattering: the scattered waves interfere to produce a single resultant wave in a given direction.

Colligative: depending upon the number of molecules, not on their identities.

Complexome: all the protein complexes of a cell.

Concatemer: two molecules linked together topologically.

Configuration: three-dimensional arrangement of atoms at the chiral center of a molecule.

Conformation: the three-dimensional structure of a large molecule defined by rotations about covalent bonds.

Cooperativity: phenomenon by which one event on a molecule increases or decreases the probability of further such events.

Corepressor: small molecule that increases the affinity of a repressor for its operator.

Co-solute, co-solvent: additional compounds, such as salts or denaturants, that are added to aqueous solutions of macromolecules.

Covalent: involving the sharing of one or more pairs of electrons.

Covariation: correlated variation of two or more variables.

Cryo-: pertaining to very low temperatures.

Cryoprotectant: substance that protects against low temperatures, especially freezing.

Cryosolvent: solvent that remains liquid at very low temperatures.

Denaturant: reagent that causes proteins or nucleic acid molecules to unfold.

Dialysis: adding or removing small molecules from a solution by their diffusion across a semipermeable membrane.

Diamagnetic: having only paired electrons and a negative magnetic susceptibility; diamagnetic substances move out of magnetic fields.

Diastereomers: molecules with different chirality.

Diffraction: scattering of radiation from atoms or molecules organized in an ordered array.

Diffusion: the spontaneous movement of molecules due to their kinetic energy.

Dipole: separation of charge within a molecule.

Diprotic: having two acid groups.

Disulfide: two sulfur atoms linked by a single covalent bond.

Eclipsed: being behind another.

Elastic scattering: the scattered beam has the same energy as the incident beam.

Electrolyte: consisting of ions.

Electronegative: attracting electrons.

Electrophile: molecule or group that is electron-deficient and reacts with nucleophiles.

Electrophilic catalysis: increase in rate of a reaction by stabilization of a negative charge that develops in its transition state.

Electrophoresis: movement of molecules or particles under the influence of an electric field.

Ellipticity: difference in absorbance of left- and right-circularly polarized components of plane-polarized radiation, measured in circular dichroism.

Empirical: based on experiment and observation, rather than theoretical.

Enantiomer: one of two isomers of a chiral compound.

Endo-: acting on the interior residues of a polymer.

Enhancer: transcription factor required for the expression of a gene.

Enzyme: protein that catalyzes a chemical reaction.

Epitope: Sites on a molecule recognized directly by an antibody.

Equilibrium: state of a chemical reaction in which the forward and backward rates are equal, so there is no net change in the concentrations of the reactants and products.

Eukaryote: organism whose cells contain a true nucleus; all organisms other than viruses, bacteria, and blue-green algae.

Exciton: high-energy excited state of a molecule or array of molecules.

Exo-: acting on the terminal residues of a polymer.

Exocytosis: release from the interior of a cell.

Exon: segment of a gene that is present in the mature messenger RNA and used in translation.

Extensive property: one that depends upon the size of the system (e.g. mass, volume).

Fatty acid: long-chain aliphatic carboxylic acid normally found esterified to glycerol.

Fluorescence: emission of light from molecules in excited electronic states.

Fluorescence lifetime: average amount of time between absorption of light and emission as fluorescence.

Fluorography: detection of radioactivity by the fluorescence emitted by a scintillator in close contact.

Fluorophore: fluorescent molecule or moiety.

Free radical: molecule containing one or more unpaired electrons.

Gene: the basic unit of genetic information, usually a segment of DNA whose expression results in the production of a messenger RNA that most often is translated into a protein.

Genetic: information present in genes.

Genetic code: the way in which the codons of a messenger RNA are read during translation and formation of a polypeptide chain.

Genetics: mechanisms by which the genetic information is transferred from one generation to the next.

Genome: the DNA and genes contained in the whole set of chromosomes present in a cell.

Genotype: the genetic constitution of an organism.

Globular: having a compact folded molecular structure.

Glycolipid: sugar linked to one or more fatty acyl groups.

Glycoprotein: protein with carbohydrate units attached covalently.

Glycosidase: enzyme that hydrolyzes glycosidic links in carbohydrate polymers.

Half-life: time taken for radioactivity or a reactant to decrease to half its original value.

Hapten: a small molecule that mimics part or all of the antigenic site of a larger molecule and interferes with its binding to an antibody.

Hard ion: one that is small, compact and not readily polarized and tends to interact with other hard ions.

Heat capacity: that quantity of heat required to increase the temperature of a system or substance by one degree centigrade or Kelvin.

Heteroduplex: a double-stranded nucleic acid molecule in which the two strands are not identical.

Heterotropic: involving different molecules.

Heuristic: involving observation and trial-and-error methods.

Histone: small basic protein that, with DNA, forms the nucleosome.

Holoenzyme: enzyme including its coenzyme.

Homeostasis: relatively stable system of interdependent elements.

Homologous: proteins or nucleic acids that arose from a common evolutionary ancestor and consequently have related sequences.

Homotropic: involving the same type of molecules.

Hormone: a chemical, nonnutrient, intercellular messenger that is effective at very low concentrations.

Hybridization: annealing complementary strands of nucleic acid.

Hydration: association with water.

Hydrogen bond: a noncovalent bond between a hydrogen donor and acceptor.

Hydrolysis: breaking a covalent bond by reaction with water.

Hydrophilic: attracted to water.

Hydrophobic: not attracted to water, but to nonpolar environments.

Hypertonic: having a high osmotic pressure.

Hypervariable region: a segment of genomic DNA characterized by considerable variation in the number of tandem repeats or a high degree of polymorphism due to point mutations.

Hyphenate: to extend one technique by combining it with another, such as LC-MS (liquid chromatography combined with mass spectrometry).

Hypotonic: having a low osmotic pressure.

Hysteresis: when the forward and reverse processes follow different paths, due to slow equilibration of the system.

Immunogen: substance that elicits an antibody response.

In silico: performed with a computer.

In situ: in position.

In vacuo: in a vacuum.

In vitro: in the test-tube.

In vivo: in the living organism.

Incoherent scattering: scattered radiation that is the sum of the individual scattered waves, with no interactions between them.

Inducer: small molecule that decreases the affinity of a repressor for its operator.

Inelastic scattering: the scattered beam has either greater or lesser energy than the incident beam, having exchanged energy with the scatterer.

Intensive property: one that is independent of the size of the system (e.g. temperature).

Intron: segment of a gene that is removed from the messenger RNA before it is translated.

Ionic: having a net charge.

Ionizing radiation: photons or sub-atomic particles with sufficient energy to produce ionization events while passing through matter.

Isoelectric point: pH at which a molecule exhibits no net charge and does not migrate in an electric field.

Isoionic point: the pH of a solution containing only the macromolecule of interest from which all other ions, except for H^+ and OH^-, have been removed.

Isomers: compounds with the same molecular formula but differing in the nature or sequence of bonding of their atoms, or in their spatial arrangement.

Isothermal: at constant temperature.

Isotonic: having the same osmotic pressure.

Isotopes: atoms with the same atomic number (protons) but different mass numbers (protons plus neutrons)

Isotropic: exhibiting properties with the same values when measured along axes in all directions.

Isozyme, isoenzyme: enzyme that is closely related in sequence, structure and activity to another enzyme, usually from the same source.

Kinase: enzyme that transfers a phosphate group to a protein, usually from ATP.

Le Chatelier principle: to every action there is an equal and opposite reaction.

Lectin: protein other than an antibody that recognizes specific polysaccharides.

Lewis acid: atom having empty *d* electron orbitals that act as electron sinks.

Ligase: enzyme that covalently join the ends of nucleic acids through a phosphodiester bond.

Lipid: molecule soluble in nonpolar organic solvents.

Lipoprotein: complex of proteins (apolipoproteins) and lipids.

Lone pair: pair of valence electrons that are not involved in covalent bond formation.

Lyophilization: removal of volatile solvent by subjecting a frozen solution to a vacuum, so that the solvent sublimes but any nonvolative materials are left behind.

Lysogeny: integration of a viral genome into that of the cell, rather than lytic multiplication.

Lytic: leading to lysis of the cell.

Macrostate: state of a system defined by its macroscopic properties.

Melting temperature: temperature at mid-point of a thermally-induced transition.

Mesophile: organism that grows optimally at normal physiological temperatures.

Messenger RNA: the RNA copy of a gene that is produced by transcription and used for translation into a polypeptide chain.

Microstate: state of a system defined by its individual molecules.

Minisatellites: regions of tandem repeats in the genome.

Mismatch: any base pair other than the normal A•T (or U) and G•C.

Molality: moles of substance per 1000 g of solvent

Molarity: moles of substance in 1000 ml of solution

Mole: mass in grams of the molecular weight of a molecule, containing N_A molecules.

Molecular weight: the sum of the atomic weights of all the atoms in a molecule.

Molecule: the smallest unit of matter that can exist by itself and retain all the properties of the original substance.

Monochromatic: composed of a single wavelength.

Monoclonal antibodies: homogeneous antibodies synthesized by a population of identical antibody-producing cells.

Monodisperse: homogeneous population of identical molecules

Morphology: external shape adopted by a solid.

Mutation: a change in the structure of the genome DNA, usually of the nucleotide sequence, that is passed on to future generations.

Nascent: newly-synthesized.

Negative stain: visualizing a structure by observing the shell that it leaves in an amorphous solid medium.

Nonpolar: having no functional or reactive groups, only inert hydrocarbons.

Nuclease: enzyme that hydrolyzes phosphodiester bonds of nucleic acids.

Nucleophile: molecule or group that is electron-rich and reacts with electrophiles.

Nucleosome: fundamental unit of chromatin, consisting of histones and about 200 base pairs of double-stranded DNA.

Nucleotide: monomer that upon polymerization generates a nucleic acid, either DNA or RNA.

Oligonucleotide: a short polynucleotide, usually 2 to 20 nucleotides in length.

Oligosaccharide: linear or branched carbohydrate consisting of 2 to 20 monosaccharides, linked by glycoside bonds.

Operator: region on a DNA molecule upstream of a gene at which a repressor binds and blocks transcription.

Operon: several linked genes subject to the same control.

Ordinate: the vertical axis of a graph.

Osmolyte: molecule that contributes significantly to the osmotic pressure.

Osmosis: net movement of molecules through a semipermeable membrane.

Osmotic pressure: pressure required to stop osmosis.

Oxidase: enzyme that catalyzes an oxidation using O_2 as the electron acceptor, without incorporating the O atoms into the product.

Oxidation: a chemical reaction that removes electrons, often by transferring them to O_2 to produce water.

Oxidoreductase: enzyme that catalyzes a reaction involving electron transfer to or from an external electron carrier, usually a redox protein.

Oxygenase: enzyme that catalyzes the reaction between O_2 and an organic substrate, adding O atoms to the substrate.

Palindrome: nucleotide sequence from the 5′-end to the 3′-end that is the same as its complement on the other strand.

Paramagnetic: having unpaired electron spins; a paramagnetic substance tends to move into a magnetic field.

Peptidase: enzyme that cleaves the peptide bonds of peptides or small proteins.

Peptide: a short linear segment of amino acids linked by peptide bonds.

Peptide bond: covalent bond between the α-amino and α-carboxyl groups of two amino acids.

pH: negative logarithm of the hydrogen ion activity or concentration.

Phage, bacteriophage: a virus that replicates in bacteria.

Phenotype: the observable properties of an organism, resulting from the interaction of the genotype and environment.

Phospholipid: any lipid containing phosphate, usually referring to lipids based on 1,2-diacylglycero-3-phosphate.

Phylogeny: pathway by which genes, nucleic acids, proteins, individuals, species, or populations arose and diverged during evolution.

Piezoelectric: changing its shape in response to an electric fi eld.

pK_a: the pH at which a polar chemical group is half ionized.

Plasmid: DNA molecule that is stably inherited genetically without becoming part of a chromosome.

Plectonemic: interwound.

Polar: having functional, reactive, and ionizable groups.

Polarized light: light that exhibits different properties in different directions at right angles to its direction.

Polarizer: instrument that polarizes light.

Polyamide: repeated amide (-CO-NH-) units, usually formed by polymerization of amino and carboxyl groups.

Polychromatic: comprised of many colors.

Polymer: compound formed by polymerization and consisting essentially of repeating structural units.

Polymerase: enzyme that catalyzes polymerization.

Polymorphism: ability to assume different forms.

Polynucleotide: a linear polymer produced by condensation of nucleotides.

Post-transcriptional: occurring after biosynthesis of an RNA molecule.

Post-translational: occurring after biosynthesis of a polypeptide chain.

Precess: to undergo a relatively slow gyration of the rotation axis of a spinning body about another line intersecting it so as to describe a cone.

Primary structure: sequence of amino acid or nucleotide residues in a protein or nucleic acid.

Probe: a labelled oligonucleotide used to detect complementary sequences in DNA or RNA.

Prokaryote: organism that lacks a true nucleus; a bacterium, virus, or blue-green algae.

Promoter: region on a DNA molecule upstream of a gene at which an RNA polymerase binds and initiates transcription.

Prosthetic group: any chemical group of a protein that was not part of the primary structure but acquired by binding another molecule.

Proteinase (protease): enzyme that hydrolyzes peptide bonds.

Proteome: all the proteins produced by the cell.

Pseudoknot: highly structured RNA secondary structural motif.

Quaternary structure: involving aggregation of two or more individual protein or nucleic acid molecules.

Radian: the angle subtended by an arc of a circle equal to its radius (57°).

Radical ion: free radical with a positive or negative charge.

Radioactivity: emission of ionizing radiation by atoms.

Radioisotopes: isotopes that have unstable nuclei and decay to a stable state by the emission of ionizing radiation.

Random coil: a flexible polymer in which the conformational properties of each residue are independent of those of other residues not close in the covalent structure.

Receptor: structure on a cell that binds other molecules specifically and produces a response.

Redox protein: protein that can exist reversibly in more than one oxidation state; usually it has a cofactor that handles the electrons.

Reduction: a chemical reaction that adds electrons, often by transferring H atoms.

Relaxation time: time required for a change to reach 1/e (0.368) the final state; the reciprocal of the rate constant for the reaction.

Replication: copying of DNA or RNA molecules to make multiple identical copies.

Repressor: a protein that binds to an operator to prevent transcription of a gene.

Reptating: a linear polymer moving in a snake-like manner.

Residue: individual amino acid unit of a polypeptide chain.

Resonance: two or more alternative electronic structures are required to describe a molecule.

Restriction enzyme: enzyme that recognizes a specific sequence of DNA and cleaves the backbone at or near this sequence.

Reverse transcriptase: enzyme that uses a single-stranded RNA molecule as template to synthesize a complementary strand of DNA.

Reverse transcription: the synthesis of a complementary DNA molecule from an RNA template, catalyzed by a reverse transcriptase.

Ribozyme: RNA molecule with catalytic ability.

Root-mean-square: the square root of the average value of the squares of the individual values, weighted by the probability of that value occurring.

Secondary structure: local conformation adopted by interactions only between residues close in the sequence of a protein or nucleic acid.

Sedimentation: movement of molecules or particles in a gravitational field.

Semipermeable: permeable to some molecules but not others.

Singlet state: having zero electronic spin.

Soft ion: one that is large and relatively polarizable and tends to interact with other soft ions.

Solute: that constituent of a solution that is considered to be dissolved in the other, the solvent.

Spin: intrinsic angular momentum of a nucleus or an unpaired electron that induces magnetic momentum.

Splicing: removal of introns from a messenger RNA precursor.

Stereoisomers: isomers differing only in the spatial arrangement of their atoms.

Sticky end: single-stranded nucleotides protruding from the end of a double-stranded nucleic acid, which may hybridize with another single-stranded nucleic acid with a complementary sequence.

Stochastic: random.

Substrate: the specific compound on which an enzyme acts.

Tandem repeat: end-to-end duplication of a series of identical or almost identical segments of DNA (usually of 2 to 80 base pairs).

Tautomers: isomers that are readily interconverted spontaneously and normally exist together in equilibrium.

Tertiary structure: the folded conformation adopted by a substantial segment of polypeptide or oligonucleotide chain involving interactions between groups distant in the covalent structure.

Tetrahedral: in the shape of a regular tetrahedron with four identical faces.

Torus: a shape like a donut, with a hole in the middle.

Trans-acting: acting on other molecules.

Transcription: expression of the nucleotide sequence of a gene into a messenger RNA with a complementary base sequence.

Transcription factor: regulatory protein that binds to a promoter or to a nearby sequence of DNA to facilitate or prevent initiation of transcription.

Transition metal: element with incompletely filled *d* subshell of electrons or that gives rise to cations with incompletely filled *d* subshells.

Transition state: the least stable species that occurs during a chemical reaction; its free energy determines the rate of the reaction.

Translation: expression of the genetic information of a messenger RNA into a polypeptide chain.

Triplet state: an atom or molecule with total spin quantum number of one.

Tunneling: quantum-mechanical phenomenon by which a particle can move from one energy state to another by penetrating, rather than traversing, an energy barrier as a wave.

Valence electron: outer electrons of an atom that are involved in forming chemical bonds.

Vicinal: adjacent sites in a molecule.

Virus: a structure comprised of proteins and nucleic acids that can infect a host cell and replicate to produce many more such structures.

Zwitterion: molecule with both positively- and negatively-charged groups.

~ CHAPTER 1 ~

THERMODYNAMICS FOR MOLECULAR BIOLOGY

Thermodynamics, with its discussions of Carnot engines and the pressures of nonideal gases, is undoubtedly the subject least likely to set a molecular biologist's heart racing. Yet aspects of thermodynamics are crucial for molecular biology. Will my piece of DNA adopt the A-, B- or Z-form under my conditions? Is this enzymatic reaction likely to require ATP or GTP? Will this DNA polymerase be suitable for use in PCR? All of these questions can be answered if the free energies and enthalpies of the molecules involved are known. Thermodynamics describes the energies of molecules and systems and how the equilibria of chemical processes depend upon the conditions, especially the temperature. It is a vast and mature subject, but only a few aspects of it are pertinent to molecular biology, where the pressure is generally kept constant at one atmosphere and most reactions take place in dilute aqueous solution containing salts and buffer molecules. Only these pertinent aspects of thermodynamics will be described here.

1.1. EQUILIBRIUM CONSTANTS

The most fundamental thermodynamic parameter for a chemical process or reaction is its equilibrium constant. In the case of a unimolecular reaction, such as an isomerization $A \leftrightarrow P$, it can be expressed as:

$$K_{eq} = \frac{[P]_{eq}}{[A]_{eq}} \tag{1.1}$$

where the brackets indicate the equilibrium concentration of that species. **In this and subsequent equations in this chapter, the concentrations are understood to be those at equilibrium.** Concentrations are generally expressed in terms of **molarity** (mol/liter), although mole fractions are often used with high concentrations of reactants. The value of K_{eq} can be calculated from Equation 1.1 if the molar concentrations of the reactants are known at equilibrium. Ideally, the value of this equilibrium constant will be independent of the absolute concentrations of A and P, but **nonideality is usually observed at very high concentrations**, where other interactions between the reactant molecules become significant. Such thermodynamic nonideality is usually treated by using the **activities** of the species rather than their concentrations. If the equilibrium constant calculated using activities is not independent of the absolute concentrations of A and P, there is an additional participant in the reaction that is being omitted from the equations.

Although the reactants are written in equations as just A or P, **they actually include all the other components of the solvent with which they interact.** These additional molecules may be involved in the reaction; for example, water molecules, protons or ions might be bound more by one of the reactants or products than the others and consequently would be taken up or released by the reaction. They need not be included explicitly in the reaction so long as their bulk concentrations are kept constant, but they do if the equilibrium constant is observed to change when their concentrations are varied. In molecular biology and biochemistry, reactions involving uptake or release of water molecules or hydrogen ions are extremely common but these species are generally not included in the equilibrium constant: the concentration of water is taken as unity (mole fraction) and the pH is taken to be 7. In these cases, the simplified equilibrium constants and thermodynamic parameters are usually indicated with a prime (e.g. K_{eq}'). If an equilibrium constant is found to depend upon the pH or low concentrations of salt, for example, protons or ions are involved in the reaction and need to be considered explicitly if the pH or salt concentration is varied. This is readily accomplished by expanding the equation for the reaction and for the equilibrium constant to include all relevant species. In general, a reaction involving multiple participants is written as:

$$mA + nB + oC \leftrightarrow xP + yQ + zR \tag{1.2}$$

where *m*, *n*, *o*, *x*, *y* and *z* are the numbers of molecules of each participant. The equilibrium constant is then expressed as:

$$K_{eq} = [P]^x [Q]^y [R]^z / [A]^m [B]^n [C]^o \tag{1.3}$$

where the concentration of each reactant is raised to the power of the number of molecules participating in the reaction.

One example of a reaction involving hydrogen ions is ionization of a molecule:

$$\begin{array}{ccc} A + B & \rightleftharpoons & C + HD \\ & & \updownarrow K_i \\ & & D^- + H^+ \end{array} \tag{1.4}$$

where

$$K_i = \frac{[H^+][D^-]}{[HD]}$$

is the ionization constant. In this case, the observed value of the equilibrium constant will depend upon the pH, which is the negative logarithm of the hydrogen ion concentration (pH = –log $[H^+]$). The apparent equilibrium constant K_{app}, in which HD and D^- are not distinguished, will depend upon the pH according to:

$$K_{app} = \frac{[C]([HD]+[D^-])}{[A][B]} = K_I(1 + \frac{K_i}{[H^+]}) \tag{1.5}$$

where K_l is the equilibrium constant at low pH values (where all of the product D is in the form DH). This equation demonstrates that ionization of DH at high pH (lower concentration of H^+) pulls the reaction to the *right*. If more than one ionizable species participates in the reaction, Equations 1.4 and 1.5 are correspondingly more complex but can be analyzed in the same way.

At equilibrium, the concentrations of the reactants do not change with time, so long as the conditions are not changed. The system is still dynamic, but the rates of the forward and reverse reactions are equal, so the concentrations of the reactants and products remain constant. The value of K_{eq} may therefore also be calculated from the ratio of the rate constants (Chapter 4) for the forward and reverse reactions, k_f and k_r, respectively:

$$k_f\,[A]_{eq} = k_r\,[P]_{eq} \tag{1.6}$$

$$k_f/k_r = [P]_{eq}/[A]_{eq} = K_{eq} \tag{1.7}$$

$$K_{eq} = k_f/k_r \tag{1.8}$$

1.2. GIBBS FREE ENERGY

If reactants A and P are present at the same concentration at equilibrium, they are considered to have the same **energy**. If one predominates at a higher equilibrium concentration than the other, it is the **more stable** and is said to have a **lower** energy than the other. When the pressure is kept constant, as in most molecular biology experiments, the relevant energy is the **Gibbs free energy**, normally abbreviated as G. The alternative is the Helmholtz free energy, which applies when the volume is kept constant (which will require changes in the pressure if the temperature changes). Most biochemical experiments are carried out at constant pressure, so molecular biologists almost invariably deal with the Gibbs free energy.

In most cases, molecular biologists are interested in relative, not absolute, free energies. The difference in free energies of the reactants and products, the **standard free energy change** ($\Delta G°$) per mole of reactant, is given by the natural logarithm of the equilibrium constant, multiplied by the gas constant (R) and the absolute temperature (T):

$$\Delta G°_{P-A} = -RT \log_e K_{eq} = -2.303\ RT \log_{10} K_{eq} = G_P - G_A \tag{1.9}$$

The factor 2.303 relates the two logarithmic scales to base *e* (2.718) or to base 10. The temperature in all thermodynamic equations is the absolute temperature K in units of *kelvins* (K = ° C + 273.15). The logarithmic dependence of Equation 1.9 follows from the probability of molecules existing in a specific state that is described by statistical thermodynamics.

Conversely, the equilibrium constant can be expressed in terms of the free energy difference:

$$K_{eq} = \exp(-\Delta G°_{P-A}/RT) \tag{1.10}$$

At 25° C, the value of 2.303 RT is 1.36 kcal/mol (5.69 kJ/mol). Therefore an equilibrium constant of 10 implies that the product has a free energy that is 1.36 kcal/mol (or 5.69 kJ/mol) lower than that of the reactant. An equilibrium constant of 10^2, or 100, results if the free energy difference is twice as great.*

These considerations also apply to reactions involving multiple reactants, for example:

$$A + B \leftrightarrow P + Q \quad (1.11)$$

$$K_{eq} = [P]_{eq}\,[Q]_{eq}/[A]_{eq}\,[B]_{eq} \quad (1.12)$$

$$\Delta G° = -RT \log_e K_{eq} = G_P + G_Q - G_A - G_B \quad (1.13)$$

In this case, however, the equilibrium constant and ΔG° do not provide the relative free energies of individual reactants but of all of them collectively.

If the numbers of reactants and products are not equal, the equilibrium constant will have some dimension of concentration (Equation 1.3) and it is not strictly correct to take the logarithm of a dimensioned number. In this case, an equilibrium constant with dimensions is referred to some fixed concentration, known as the **standard state,** which is usually taken for simplicity as either 1 mol/liter (1 M) for a solute or 1 mole fraction for a solvent, such as water. It must be remembered that **all the thermodynamic quantities calculated from such a dimensioned equilibrium constant will pertain only to the standard state concentration of reactants.**

Other components of the system can be involved in the reaction. For example, a reaction in dilute aqueous solution can release or take up *n* water molecules if the reactants and the products differ in their affinities for water:

$$A + B \leftrightarrow P + Q + nH_2O \quad (1.14)$$

$$K_{eq} = [P]\,[Q]\,[H_2O]^n/[A]\,[B] \quad (1.15)$$

$$\Delta G° = -RT \log_e K_{eq} = -RT \log_e ([P]\,[Q]\,[H_2O]^n/[A]\,[B]) \quad (1.16)$$

In this case, *n* need not be an integer and can be either positive or negative. The biochemistry convention, however, defines the activity of pure water as unity:

$$\Delta G°' = -RT \log_e K_{eq}' = -RT \log_e ([P]\,[Q]/[A]\,[B]) \quad (1.17)$$

and

$$\Delta G° = \Delta G°' + nRT \log_e ([H_2O]/\text{M}) \quad (1.18)$$

* *Energies should strictly be expressed in SI units of joules (J), but calories (cal) are still widely used in molecular biology, so both will be used here (1 cal = 4.184 J).*

Because $[H_2O]$ is normally 55.5 M, usually:

$$\Delta G^\circ = \Delta G^{\circ\prime} + 5.46\, n \text{ kcal/mol} \tag{1.19}$$

The standard free energy change for a reaction at equilibrium can be used to determine whether any reaction will occur, and in which direction, starting with any particular initial concentrations of the reactants. For example, if the initial concentrations of the reactants and products of Equation 1.2 are indicated by the subscript *i*, the actual change in free energy for the spontaneous reaction, ΔG, will be given by:

$$\Delta G = \Delta G^\circ + RT \log_e([P]_i^x\,[Q]_i^y\,[R]_i^z/[A]_i^m\,[B]_i^n\,[C]_i^o) \tag{1.20}$$

The reaction of Equation 1.2 will proceed from *left* to *right* only if the value of ΔG is negative. If ΔG is positive, the reaction will proceed in the reverse direction. At equilibrium, $\Delta G = 0$.

Note the very fundamental difference between ΔG and ΔG°. Do not confuse the two.

Calculation of standard transformed Gibbs energies and standard transformed enthalpies of biochemical reactants. R. A. Alberty (1998) *Arch. Biochem. Biophys.* **353**, 116–130.

Standard apparent reduction potentials of biochemical half reactions and thermodynamic data on the species involved. R. A. Alberty (2004) *Biophys. Chem.* **111**, 115–122.

1.2.A. Coupled Reactions

An energetically unfavorable reaction, for which ΔG is positive, can be made to occur by coupling it to another reaction with a more negative ΔG. For example, consider the two-step reaction process:

$$A + B \leftrightarrow C + D\ (\Delta G_1^\circ > 0) \tag{1.21}$$

$$D + E \leftrightarrow F + G\ (\Delta G_2^\circ < 0) \tag{1.22}$$

The first reaction (Equation 1.21) is not favored energetically and will not proceed spontaneously very far starting with just A and B, because $\Delta G_1^\circ > 0$. It will, however, produce a small amount of the product D, which can then be used in the energetically favorable second reaction (Equation 1.22). If $\Delta G_1^\circ + \Delta G_2^\circ < 0$, the equilibrium concentration of D in the second reaction will be lower than that in the first, so it will be consumed in the second reaction. This can also be illustrated with the overall reaction:

$$A + B + E \leftrightarrow C + F + G\ (\Delta G_1^\circ + \Delta G_2^\circ < 0) \tag{1.23}$$

The second reaction will therefore drive (or pull) the first, and the two reactions are said to be coupled through their common intermediate D.

In general with multi-step reactions, like that of Equation 1.23, the equilibrium constant for the overall reaction is the product of the equilibrium constants of the individual steps. The change in free energy is the sum of the values for the individual steps.

Such coupling of reactions is the way that most biosynthetic reactions are driven in biological systems. Coupling two amino acids in a peptide bond, or two nucleotides in a phosphodiester bond, is not energetically favorable, but both reactions are driven during biosynthesis by coupling them to the energetically highly favorable hydrolysis of ATP.

Thermodynamics and bioenergetics. Y. Demirel & S. I. Sandler (2002) *Biophys. Chem.* **97**, 87–111.

1.2.B. Linked Functions

The physical law of conservation of energy requires that there be no net change of any energetic parameter in any cyclic process. Consequently, any chemical process that can be written as a cycle must have no net change in energy around the cycle, irrespective of which reaction path is followed. The sum of the changes in free energy, enthalpy and entropy around any such cycle must be zero, and the product of the equilibrium constants must be unity. This has a most important consequence for macromolecules, in that any two phenomena on such a macromolecule that affect each other, such as the binding of two different ligand molecules to the same macromolecule, must have equal and identical effects on each other. Such effects are known as **linked functions**. For example, consider the occurrence of two phenomena A and B that can occur simultaneously to the macromolecule, P, with the indicated equilibrium constants:

Equilibrium constant K^A pertains to phenomenon A in the absence of B, whereas $K^{A/B}$ pertains to phenomenon A in the presence of B; similarly for phenomenon B in the absence and presence of A. Because the free energy change around such a cycle must be zero and the products of the equilibrium constants must be unity:

$$\frac{K^A K^{B/A}}{K^B K^{A/B}} = 1 \tag{1.25}$$

$$\frac{K^A}{K^{A/B}} = \frac{K^B}{K^{B/A}} \tag{1.26}$$

In other words, **whatever effect phenomenon A has on the equilibrium constant for phenomenon B, the occurrence of B must have exactly the same effect on the equilibrium constant for phenomenon A.**

Linkage relationships like this are most commonly encountered for the binding of small molecules by proteins. For example, the affinity of hemoglobin for oxygen is affected by the binding of organic phosphates, hydrogen ions and CO_2 at other sites on the hemoglobin molecule; likewise, the binding of oxygen must affect the affinity of hemoglobin for these other ligands.

Linkage relationships like this apply not just to ligand binding but also to any two phenomena that affect each other in a macromolecule. For example, a disulfide bond, or any other interaction, that stabilizes a particular protein conformation must be stabilized to the same extent by the presence of that conformation. If one interaction in a folded conformation affects the stability of that folded conformation, and of the other interactions that stabilize that conformation, the first interaction must be affected to the same extent by the other interactions. If the stability of a conformation is altered by changes in the pH, the ionization of the relevant groups must be affected to the same extent by the presence of that conformation. The importance of linked functions for understanding macromolecules cannot be overly emphasized.

Linkage graphs: a study in the thermodynamics of macromolecules. J. Wyman (1984) *Quart. Rev. Biophys.* **17**, 453–488.

Binding and Linkage. J. Wyman & S. J. Gill (1990) University Science Books, Mill Valley, CA.

1.3. ENTHALPY

The free energy is comprised of the **enthalpy** (H) and the **entropy** (S) according to the basic equation:

$$\mathbf{G = H - T\,S} \qquad (1.27)$$

The entropy S is multiplied by the temperature (T), whereas the enthalpy is not. Alternatively, Equation 1.27 can be written as:

$$G/T = H/T - S \qquad (1.28)$$

In terms of the equilibrium constant, combining Equations 1.9 and 1.27 gives:

$$\log_e K_{eq} = -\Delta H^\circ_{P-A}/RT + \Delta S^\circ_{P-A}/R \qquad (1.29)$$

The enthalpy is a measure of the total heat content of a system. The enthalpy of a molecule can be considered to be composed of (1) the translational, vibrational and rotational energies of the molecule, (2) the energy involved in its covalent bonds, (3) the energy involved in any noncovalent interactions with other molecules present, plus (4) the product of the volume and pressure of the system. The last contribution usually remains constant in biochemical reactions, when it will not contribute to changes in the enthalpy. Each of the first three factors contributes to making the enthalpy of the system more negative (i.e. more favorable energetically).

The change in enthalpy of a reaction can be measured experimentally (in a calorimeter, Section 1.6) by the amount of heat that is either liberated or taken up in a reaction:

$$A \leftrightarrow P + \text{heat} \tag{1.30}$$

By considering heat as a product, it is apparent that in this case increasing the temperature will drive the reaction to the *left*. This is an **exothermic reaction**, and it occurs with a negative change in the enthalpy. The opposite, a positive value of the enthalpy change, indicates that heat is absorbed by the system, and it is an **endothermic reaction**. An increase in temperature will drive such a reaction to the *right*.

The enthalpy change is a measure of the temperature-dependence of the equilibrium constant for the reaction, and it can also be measured in this way. This is apparent by rearranging Equation 1.29 to:

$$\log_e K_{eq} = (-\Delta H^\circ_{P-A}/R)(1/T) + \Delta S^\circ/R \tag{1.31}$$

According to this equation, a plot of $\log_e K_{eq}$ versus (1/T), known as a **van't Hoff plot** (Figure 1-1), will have a slope proportional to the enthalpy change ($-\Delta H^\circ_{P-A}/R$). It will be a straight line if ΔH° and ΔS° are independent of temperature. The entropy change would be given as ΔS°/R by the intercept at 1/T = 0 (T = ∞), but measurements over a very wide temperature range would be required to make such an extrapolation credible. Fortunately, there are more accurate ways to determine ΔS°: ΔH° is known from the slope of the curve, and ΔG° is known from the equilibrium constant, so the difference between the two gives ΔS° (Equation 1.27). If ΔH° and ΔS° vary with the temperature, the slope of the line will vary, as in Figure 1-1. Such nonlinearity indicates that the two reactants have different heat capacities (Section 1.5).

Differentiating Equation 1.28 relative to T (assuming H° to be constant) yields:

$$d(\Delta G^\circ/T)/dT = -\Delta H^\circ/T^2 \tag{1.32}$$

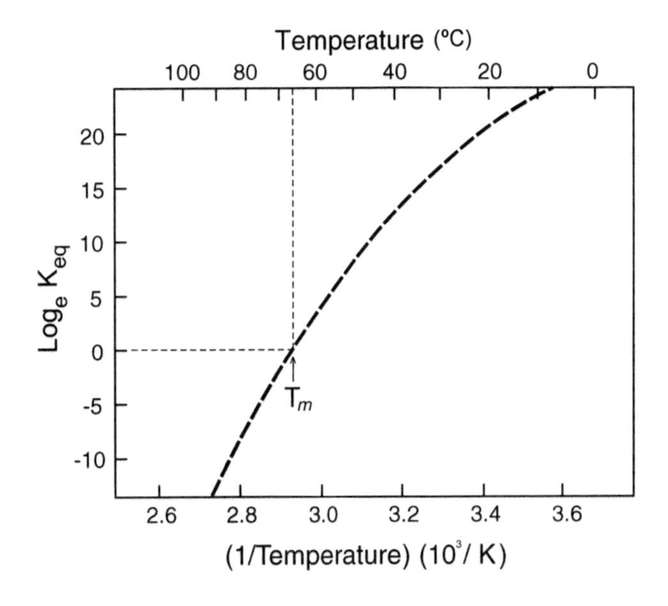

Figure 1-1. Example of a van't Hoff plot, covering the temperature range 10–90° C. The natural logarithm of the equilibrium constant K_{eq} is plotted versus the reciprocal of the absolute temperature (*bottom*). The corresponding temperatures in degrees Celsius are indicated at the *top*. The temperature where the equilibrium constant is 1 is known as the T_m. This curve describes the reversible thermal unfolding of the protein chicken egg-white lysozyme: K_{eq} = [N]/[U], where N is native, folded lysozyme at low temperatures and U is unfolded at high temperatures; data from W. Pfeil & P. L. Privalov (1976) *Biophys. Chem.* **4**, 23–50. The curve is not linear because U has a larger heat capacity than N. The T_m is at approximately 67° C.

which is known as the **Gibbs–Helmholtz equation.** This differentiation is straightforward if $\Delta H°$ and $\Delta S°$ do not vary with the temperature, but the same result is obtained if they are not constant, which occurs if the reactants have different heat capacities (Section 1.5). In this case, the differentiation yields:

$$d(\Delta G°/T)/dT = -\Delta H°/T^2 + (1/T)\ d(\Delta H°)/dT - d(\Delta S°)/dT \quad (1.32\text{-A})$$

The two last terms $(1/T)\ d(\Delta H°)/dT$ and $d(\Delta S°)/dT$ are both equal to the difference in heat capacity divided by the temperature (Equation 1.44) and to each other, so they cancel out and the result is the same as in Equation 1.32.

Equation 1.9 can be written as:

$$\log_e K_{eq} = -\Delta G°/RT \quad (1.33)$$

Differentiation with respect to T then gives:

$$d(\log_e K_{eq})/dT = (-1/R)\ (d(\Delta G°/T)/dT \quad (1.34)$$

Substituting in Equation 1.32 yields:

$$d(\log_e K_{eq})/dT = \Delta H°/RT^2 \quad (1.35)$$

which is known as the **van't Hoff equation.**

The enthalpy change measured in the ways described here is accurate only to the extent that the original assumption is true: that the reaction is two-state (Equation 1.30), with only the initial and final species present in significant quantities at equilibrium at each temperature.

van't Hoff and calorimetric enthalpies. II. Effects of linked equilibria. J. R. Horn *et al.* (2002) *Biochemistry* **41**, 7501–7507.

Ground-state enthalpies: evaluation of electronic structure approaches with emphasis on the density functional method. B. Delley (2006) *J. Phys. Chem. A* **110**, 13631–13639.

Kinetics and thermodynamics of sucrose hydrolysis from real-time enthalpy and heat capacity measurements. E. Tombari *et al.* (2007) *J. Phys. Chem. B* **111**, 496–501.

Enthalpy distribution functions for protein–DNA complexes: example of the binding of AT-hooks to target DNA. D. Poland (2007) *Biophys. Chem.* **125**, 497–507.

1.4. ENTROPY

The **entropy of a system is proportional to the variation of its free energy with temperature.** This can be demonstrated readily by differentiating Equation 1.27 with regard to T:

$$-S° = dG°/dT \quad (1.36)$$

assuming that H° and S° are constant and independent of temperature. If they do vary with temperature:

$$dG°/dT = dH°/dT - S° - T\ dS°/dT \quad (1.37)$$

As in Equation 1.32, the two terms dH°/dT and T dS°/dT are both equal to the heat capacity (Equation 1.44), so they cancel out and the result is the same as before.

In molecular terms, **the entropy of a system is related to its disorder**:

$$S° = R \log_e Z \quad (1.38)$$

where Z is the number of possible states in one mole.*

This can be illustrated by considering the equilibria in a system with many possible variations, such as an unfolded polymer that can exist in Z + 1 different conformations, C_1 to C_{Z+1}, all in rapid equilibrium and all with the same free energy, so that the equilibrium constant for each interconversion is 1:

$$C_1 \leftrightarrow C_2 \leftrightarrow C_3 \leftrightarrow C_4 \leftrightarrow \ldots \leftrightarrow C_{Z+1} \quad (1.39)$$

If Z is very large, as can happen with long polymers, each molecule in a small population will have a different conformation at any instant of time. For the polymer to adopt a specific conformation, say C^F_{Z+1}, in most of the molecules, that conformation would have to be sufficiently more stable than all the other possibilities (as a result of specific interactions stabilizing it) so that the equilibrium constant between conformation Z + 1 and each of the others would be greater than Z, the number of conformations. In this case, conformation C_{Z+1} would represent a specific conformation, C^F_{Z+1}, and all the other conformations, C_1 to C_Z, would represent the unfolded form of the molecule, C_U. Using brackets to designate the molar concentration of a species with a particular conformation, conformation C^F_{Z+1} will be present half the time (i.e. in half the molecules) when:

$$\frac{[C_{Z+1}]}{[C_1]+[C_2]+[C_3]+\ldots[C_Z]} = \frac{[C^F_{Z+1}]}{[C_U]} = 1 \quad (1.40)$$

The denominator of Equation 1.40 can be replaced by Z $[C_Z]$ because conformations C_1 to C_Z are postulated to have the same free energy and consequently will be present at equilibrium at the same concentration. This produces:

$$\frac{[C^F_{Z+1}]}{Z[C_Z]} = 1 \quad (1.41)$$

$$\frac{[C^F_{Z+1}]}{[C_Z]} = Z \quad (1.42)$$

* *In all such equations, the factor R gives the energy per mole, whereas the factor k_B (Boltzmann's constant) gives the energy per molecule: $R = N_A\ k_B$, where N_A is Avogadro's number, the number of molecules in a mole of substance.*

In this case, C_Z represents any of the Z individual conformations, which are all equivalent energetically. Consequently, conformation C^F_{Z+1} must have a free energy that is $-RT \log_e Z$ lower than each of the other conformations in order to be present half the time. As this contribution to the free energy is entropic, and is known as the **conformational entropy**, $S°_{conf}$, it will be multiplied by T (Equation 1.27), so it is given by:

$$S°_{conf} = R \log_e Z \tag{1.43}$$

which is the same as Equation 1.38. The conformational entropy stabilizes the form of the molecule with the greatest conformational freedom.

In general, **the greater the disorder of a state, the greater the stabilizing contribution of the entropy to its free energy**.

Why is it so difficult to simulate entropies, free energies, and their differences? W. P. Reinhardt *et al.* (2001) *Acc. Chem. Res.* **34**, 607–614.

NMR relaxation studies of the role of conformational entropy in protein stability and ligand binding. M. J. Stone (2001) *Acc. Chem. Res.* **34**, 379–388.

Estimating entropies from molecular dynamics simulations. C. Peter *et al.* (2004) *J. Chem. Phys.* **120**, 2652–2661.

Entropy of water in the hydration layer of major and minor grooves of DNA. B. Jana *et al.* (2006) *J. Phys. Chem. B* **110**, 19611–19618.

1.5. HEAT CAPACITY

The heat capacity is defined as the variation of the internal energy of the system with variation of the temperature. It is proportional to the number of degrees of freedom in the distribution of enthalpy states, the number of ways that the system can absorb energy without increasing the temperature. For example, all monoatomic gases have very similar and small heat capacities, because when absorbing heat they can only increase their velocities, which increases the temperature. More complex molecules have larger heat capacities, because they can change their rotational and vibrational states as well, so a greater amount of heat is required to increase their temperature. More rotational and vibrational states usually become accessible at higher temperatures, so the heat capacity generally increases with increasing temperature.

The heat capacity of a substance can be measured by the amount of heat required to raise its temperature. In fact, 1 cal was defined originally as the amount of heat required to increase the temperature by 1° C (= 1 K) of 1 g of water at 15° C. Because the heat required varies slightly, but significantly, with the original temperature of the water, 1 cal is now defined as 4.184 J. The heat capacity can be expressed as the molar quantity, per mole of substance, or per gram, when it is known as the **specific heat capacity**. The specific heat capacity is useful for comparing different substances (e.g. Table 1-1).

Table 1-1. Heat capacities of some liquids

	Liquid temperature (°C)	Specific heat capacity (cal/g/°C)
CCl_4	20	0.20
C_6H_6	20	0.41
CH_3COOH	0	0.47
CH_3COCH_3	20	0.53
$CH_3CH_2CH_2CH_3$	0	0.55
$CH_3CH_2CH_3$	0	0.58
CH_3CH_2OH	25	0.58
H_2O	15	1.00
NH_3	20	1.13

At constant pressure, the heat capacity is designated as C_p. It defines the temperature- dependence of the enthalpy, the entropy and the free energy:

$$C_p = \frac{\partial H}{\partial T} = T\frac{\partial S}{\partial T} = T\frac{\partial^2 G}{\partial T^2} = \frac{\langle H^2 \rangle}{kT^2} \tag{1.44}$$

where $\langle H^2 \rangle$ is the mean-squared fluctuation in the enthalpy. The partial differentiations indicate that all parameters other than the temperature are kept constant. The heat capacity generally decreases with decreasing temperature, and it goes to zero at absolute zero, 0 K (–273.15° C). The heat capacity can be measured directly, in a calorimeter, or it can be estimated (less accurately) from the temperature-dependence of the enthalpy, by the curvature of van't Hoff plots (Figure 1-1).

If the reactants and products of a reaction differ in their heat capacities, the changes in enthalpy and entropy during the reaction are both temperature-dependent. Note that they change in the same direction, so the changes to their contributions to the free energy (Equation 1.27) tend to cancel out. Consequently, reactions with large changes in heat capacity have changes in enthalpy and entropy that vary widely at different temperatures, but the free energy difference tends to change much less and can remain relatively constant. This phenomenon is known as **enthalpy–entropy compensation** and is due simply to the large change in heat capacity.

Water has a considerably higher heat capacity than most liquids, 1.00 cal/g/°C (Table 1-1), and consequently the enthalpy of water increases by approximately 100 cal/g between 0° C and 100° C. In contrast, ice has a specific heat capacity of no more than 0.50 cal/g/°C, and most organic liquids have similar values in the range of 0.4–0.6 cal/g/°C. Structural interpretation of the heat capacity is not usually straightforward, but the considerably greater heat capacities of polar liquids such as water and ammonia **can be rationalized by the importance of hydrogen bonding in their liquid structures**

and its temperature-dependence (Section 2.4). At low temperatures, the molecules in such liquids tend to be hydrogen bonded, which decreases their enthalpy, but at a free energy cost of decreased entropy because the molecules must be in relatively fixed positions to form hydrogen bonds between them. As the temperature is increased, the degree of hydrogen bonding decreases, permitting the molecules more freedom and increasing both the enthalpy and the entropy.

Reactions in aqueous solution often occur with large changes in the heat capacity, and they usually involve changes in the amount of nonpolar surface exposed to water. Such surfaces generally cause the surrounding water molecules to adopt more regular hydrogen-bonded arrays around them, thereby perturbing the structure, enthalpy and entropy of the water of hydration (Section 3.2.B). Large changes in the heat capacity are usually attributed to such solvation effects, but significant heat capacity changes are to be expected for any macromolecular process involving a multiplicity of cooperative weak interactions of whatever kind. For example, two complex molecules (or parts of a single macromolecule) can interact tightly, decreasing their enthalpy, but at an energetic cost of decreasing their flexibility and entropy as well. Conversely, less tight interactions produce a less favorable enthalpy change but are compensated by an increase in the flexibility and entropy.

Heat capacity effects in protein folding and ligand binding: a re-evaluation of the role of water in biomolecular thermodynamics. A. Cooper (2005) *Biophys. Chem.* **115**, 89–97.

Heat capacity in proteins. N. V. Prabhu & K. A. Sharp (2005) *Ann. Rev. Phys. Chem.* **56**, 521–548.

Heat capacity changes associated with nucleic acid folding. P. J. Mikulecky & A. L. Feig (2006) *Biopolymers* **82**, 38–58.

Generalized solvation heat capacities. D. Ben-Amotz & B. Widom (2006) *J. Phys. Chem. B* **110**, 19839–19849.

Heat capacity of associated systems. Experimental data and application of a two-state model to pure liquids and mixtures. C. A. Cerdeirina *et al.* (2007) *J. Phys. Chem. B* **111**, 1119–1128.

1.6. CALORIMETRY

The heat liberated or taken up by a reaction or physical process, i.e. the increase or decrease in the enthalpy, is measured most directly and accurately in a **calorimeter**. The most simple calorimeter consists of a thermometer in a container insulated to prevent exchanges of heat with the environment. Two liquids that react could be mixed in the container, and the initial and final temperatures measured. The change in heat would be obtained by multiplying the temperature change by the mass and specific heat capacities of the liquids; this value divided by the number of moles of reactants would yield the enthalpy of reaction. Modern microcalorimeters are much more sophisticated, requiring only small volumes of sample and having means of controlling the temperature very exactly. The changes in the sample containing the reactants are usually compared with those occurring in a reference that is identical in every way except for the molecules of interest. In many cases, the two samples are maintained at the same temperature, and the heat required to accomplish this is supplied electrically and measured very precisely. Modern instruments can measure changes due to the presence of a macromolecule that are only 10^{-6} those that occur in the reference solution.

Two types of calorimeters are used most frequently in molecular biology: (1) **isothermal titration calorimeters** (Figure 1-2), for measuring the enthalpies of interaction between two or more molecules, and (2) **differential scanning calorimeters** (Figure 1-3), in which the temperature of the sample is varied to detect any thermal unfolding or melting processes that occur in proteins or nucleic acids.

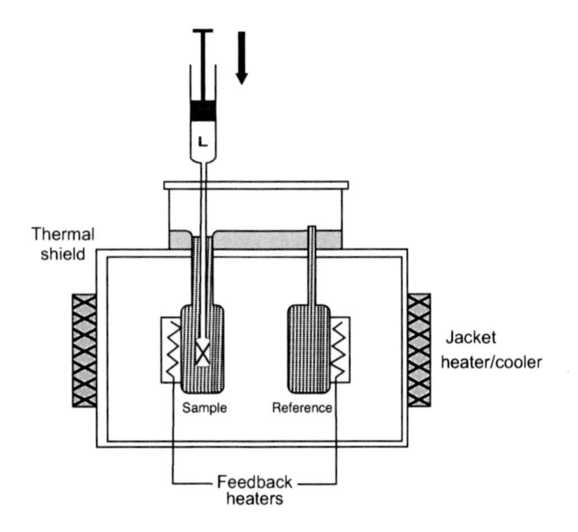

Figure 1-2. Schematic diagram of an isothermal titration calorimeter. L is the reactant that is added in small aliquots to the sample, which contains only the second reactant initially. The jacket heater/cooler is used to maintain the temperature of the entire system, whereas the feedback heaters respond to any difference in temperatures between the reference and sample liquids. Adapted from A. Cooper (2004) *Biophysical Chemistry*, Royal Society of Chemistry, Cambridge, p. 107.

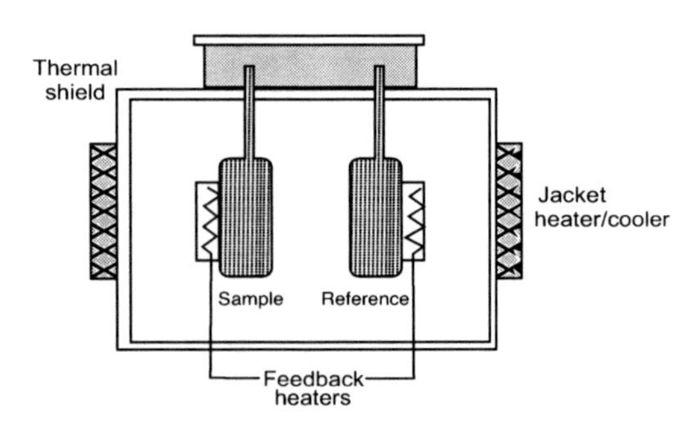

Figure 1-3. Schematic diagram of a differential scanning calorimeter. The sample contains the molecule of interest, while the reference solution is identical except for the absence of the molecule being studied. The jacket heater/cooler is used to increase the temperature of the entire system gradually, while the feedback heaters respond to any differences in temperature between the sample and reference solutions (caused by any thermal transition in one of them). These feedback heaters measure the heat capacities of the two solutions; integration of any difference between them gives the heat liberated or absorbed by the thermal transition of the sample. Adapted from A. Cooper (2004) *Biophysical Chemistry*, Royal Society of Chemistry, Cambridge, p. 104.

1.6.A. Isothermal Titration Calorimetry

The heat of the interaction between two molecules is measured by mixing them gradually. Small aliquots of one of them (reactant 2) are added stepwise to a much larger solution of the other, reactant 1 (Figure 1-2). The sample and reference solutions are identical initially, but aliquots of reactant 2 are added only to the sample. The two solutions are maintained at constant and identical temperatures using two types of heaters/coolers. One maintains the entire system at constant temperature, while the second is used to overcome any differences between the two solutions caused by the injection of reactant 2 to one of them. This second input is a measure of the heat of the reaction between them (a negative value indicates that heat has been liberated by the reaction). Of course, the appropriate measurements must be made to correct for nonspecific effects of ligand addition, stirring of the reaction mixture, etc.

Addition of each aliquot is accompanied by a burst of heat taken up or liberated that is caused by the interaction between the two reactants (Figure 1-4). As soon as thermal equilibrium has been restored, another aliquot is added. With subsequent additions, the burst of heat decreases in magnitude as reactant 1 is depleted by the reaction. When all of reactant 1 is exhausted, no further heat is liberated or released upon addition of further aliquots of 2. Summing all these events provides a curve of the extent of interaction as a function of the concentration of reactant 2, and gives the enthalpy change of the reaction (Figure 1-4, *middle*). Any change in the heat capacity can be measured by making the same measurements at different temperatures, to determine the temperature-dependence of the enthalpy change.

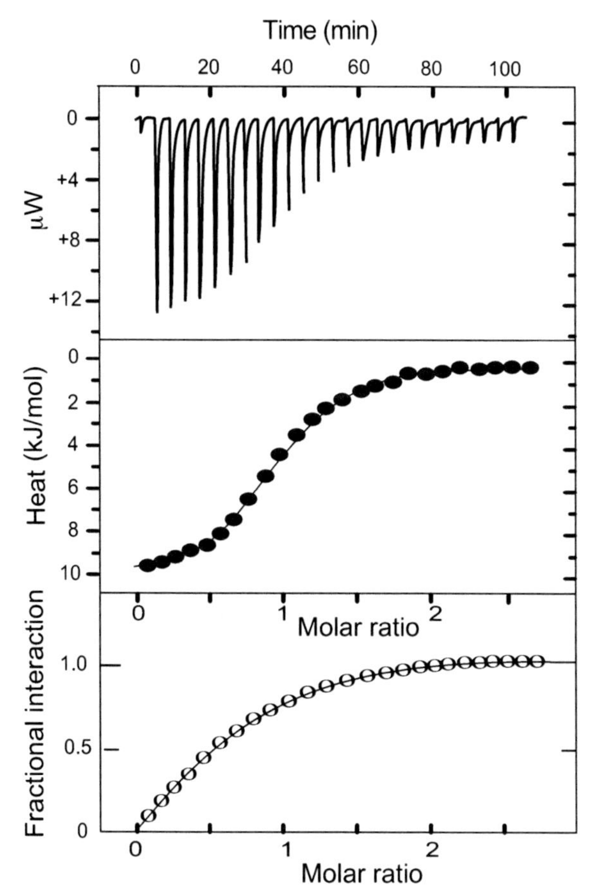

Figure 1-4. Typical results of titration of one reactant (1) in a large volume by adding small, concentrated aliquots of a second reactant (2) and measuring the heat change by isothermal titration calorimetry. The data from the isothermal titration calorimeter (Figure 1-2) are shown at the *top*. Each addition of the second reactant produces an exothermic heat pulse that is measured in watts (W) and is negative because heat is produced by the reaction. It is proportional to the extent of the reaction that takes place. This decreases as reactant 1 is used up by the reaction with reactant 2. Knowing the concentrations and volumes of the two reactant solutions, the data may be transformed to the curve given in the *middle*. The titration curve may then be constructed by integrating the original data, to give the curve at the *bottom*. The gradual change in slope of the curve indicates that the interaction between the two reactants is reversible and not complete with the concentrations used. If reactant 1 were a protein and reactant 2 a ligand that binds to it, the titration curve would give the degree of binding of ligand to the protein as a function of the total concentration of ligand added.

The reactants are usually a macromolecule (M) and a ligand (L) that binds to it. If $\Delta H^{o\prime}$ is the apparent enthalpy change associated with the formation of one mole of complex, ML, the heat produced in forming n_p moles from adding an aliquot of L to M will be:

$$Q_x = n_p \, \Delta H^{o\prime} = V \, [ML]_i \, \Delta H^{o\prime} \tag{1.45}$$

where V is the volume of the solution and $[ML]_i$ the concentration of complex generated. From the equations for binding, where K_d is the dissociation constant:

$$Q_x = V \, \Delta H^{o\prime} \, \frac{[M_{total}][L]}{K_d + [L]} \tag{1.46}$$

The ligand is added from a stock solution of concentration L_0, and each addition of ligand causes a change in the volume and therefore in the actual concentrations of all the species. Consequently, the differential heat production for the *i*th addition is given by:

$$\frac{\Delta Q_i}{\Delta[L]_i} = \frac{V_o \Delta H^{\circ\prime}([ML]_i - (1 - V_i/V_o)[ML]_{i-1} + q_d)}{[L]_i - [ML]_{i-1}} \quad (1.47)$$

where V_0 is the original volume and q_d is the heat of dilution of the ligand stock solution. The value of $[ML]_i$ can be computed for each step, since:

$$[ML]_i = \frac{([M]_i - [ML]_i)([L]_i - [ML]_i)}{K_d} \quad (1.48)$$

where the values of $[M]_i$ and $[L]_i$ can be calculated for each step. The ligand is added until there is no longer any change in the heat, indicating that the macromolecule binding sites are saturated. The experimental data for all the additions are fitted to the above equations, to yield values for M_{total}, $\Delta H^{\circ\prime}$ and K_d. In addition, these data can yield the free energy and entropy of complex formation.

Isothermal titration calorimetry. M. M. Lopez & G. I. Makhatadze (2002) *Methods Mol. Biol.* **173**, 121–126.

Three important calorimetric applications of a classic thermodynamic equation. M. J. Blandamer *et al.* (2003) *Chem. Soc. Rev.* **32**, 264–267.

1.6.B. Differential Scanning Calorimetry

The uptake or liberation of heat that occurs during some thermal transitions, such as the unfolding of a protein or the melting of a nucleic acid double helix, is measured as the temperature is gradually varied in a differential scanning calorimeter (Figure 1-3). Sample and reference solutions, usually about 1 ml in volume, are contained in identical vessels and differ only in the presence of macromolecule solely in the sample. Both solutions are heated, using the jacket heater/cooler and the main heaters, to increase their temperatures gradually (a small positive pressure is maintained to inhibit formation of bubbles by dissolved gases). In the absence of a macromolecule, the two vessels will behave identically, but its presence will affect the heat capacity. At some temperature, the macromolecules in the sample may begin to change thermally (e.g. unfold), either absorbing or producing heat. Any such transition produces a difference in temperatures of the two vessels: some of the heat energy from the main heaters will be used to bring about an endothermic unfolding transition, rather than in raising the temperature. The resulting difference in temperature will be detected and abolished by the feedback heaters. The amount of energy required for this correction is a measure of the heat of the transition.

In general, **the differences in heat energy uptake between the sample and reference cells required to maintain equal temperatures correspond to differences in their apparent heat capacities**. The thermal transition is apparent as a peak in the heat capacity difference, which reaches a maximum at the melting temperature, T_m (Figure 1-5). At T_m, the equilibrium constant for the original and final conformations has a value of 1, so the free energy change is zero. Consequently, from Equation 1.27, the enthalpy and entropy changes at this temperature are closely related:

$$\Delta H(T_m) = T_m \, \Delta S(T_m) \quad (1.49)$$

Once all the molecules are unfolded, the heat capacity curve falls to a small constant value, which may be different from the heat capacity of the folded molecules (e.g. the heat capacities of unfolded protein molecules are usually larger than those of folded molecules). **The change in enthalpy in the transition is given by the area under the heat capacity curve.**

So long as the temperature was increased sufficiently slowly that the sample was at equilibrium, **the shape of the heat capacity curve also gives information about the fraction of molecules that were folded as a function of temperature** (Figure 1-5). Assuming that only fully folded and fully unfolded molecules were present, this gives the value of the equilibrium constant for unfolding as a function of temperature, which can be used to determine the enthalpy change by van't Hoff analysis (Equation 1.31). This value should be the same as that determined by the area under the curve (the calorimetric enthalpy change) if the reaction is as simple as imagined. If not, other species were present at significant concentrations during the thermal transition.

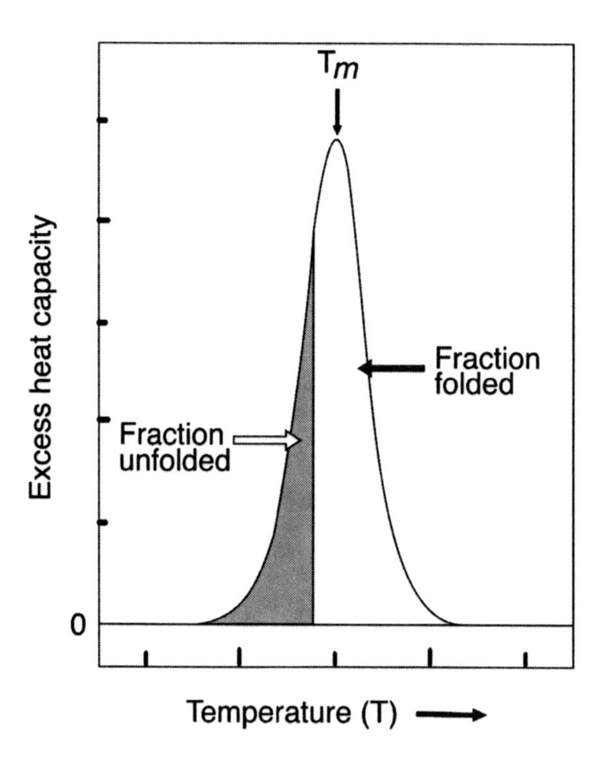

Figure 1-5. A typical heat capacity curve for a simple, one-step thermal unfolding transition at equilibrium measured by differential scanning calorimetry. The thermal transition is apparent as a peak of heat capacity, with the maximum at the mid-point of the thermal transition (the melting temperature, T_m). The area under the curve gives the enthalpy change of the transition. The shape of the curve also gives information about the fraction of molecules unfolded at each temperature, indicated by the dark area; assuming a two-state reaction, the area to the *left* is proportional to the number of molecules unfolded, whereas that to the *right* is proportional to those that are still folded.

Differential scanning calorimetry. M. M. Lopez & G. I. Makhatadze (2002) *Methods Mol. Biol.* **173**, 113–119.

Differential scanning calorimetry in life science: thermodynamics, stability, molecular recognition and application in drug design. G. Bruylants *et al.* (2005) *Curr. Med. Chem.* **12**, 2011–2020.

Advances in the analysis of conformational transitions in peptides using differential scanning calorimetry. W. W. Streicher & G. I. Makhatadze (2007) *Methods Mol. Biol.* **350**, 105–113.

~ CHAPTER 2 ~

NONCOVALENT INTERACTIONS BETWEEN ATOMS AND MOLECULES

Life is dependent upon physical interactions between the many different molecules present in a cell: water, salts, membranes, proteins, nucleic acids and the numerous other large and small molecules present in living systems. All of these interactions arise from a limited set of fundamental noncovalent forces, but with many variations on the theme, so it is important to understand their physical basis at least qualitatively. This chapter will describe briefly, and in simple terms, the types of noncovalent interactions that exist between atoms in isolation.

Probing the relation between force – lifetime – and chemistry in single molecular bonds. E. Evans (2001) *Ann. Rev. Biophys. Biomolec. Structure* **30**, 105–128.

Physical Biochemistry for the Biosciences. R. Chang (2005) University Science Books, Sausalito, CA.

Physical Chemistry for the Life Sciences. P. Atkins & J. de Paula (2006) Oxford University Press, Oxford.

2.1. SHORT-RANGE REPULSIONS: DEFINING ATOMIC VOLUME

The most important interaction, energetically and structurally, between atoms and molecules is the repulsion that eventually takes place between them as they approach each other. As they come sufficiently near for their electron orbitals to overlap, the repulsion increases enormously because the electrons on different molecules cannot be in the same part of space at the same time, as stated by the **Pauli exclusion principle**. The repulsive energy is often taken to increase with the inverse 12th power of the distance between the centers of the two atoms. A more realistic description has the energy varying exponentially with the inverse of the distance, but there is little practical difference between the two descriptions.

Because the repulsive energy rises so steeply, **it is possible to consider atoms and molecules as having definite dimensions and occupying volumes that are impenetrable to other molecules at ordinary temperatures, even though their exterior is simply an electron cloud.** Individual atoms are usually approximated as spheres, and the impenetrable volume is usually defined by the **van der Waals radius**. Values of the van der Waals radii are usually measured by the smallest distances that can exist between the centers of atoms in the crystalline state that are neighboring, but not covalently bonded; this distance is taken to be the sum of the van der Waals radii of the two atoms. Some typical values of van der Waals radii are given in Table 2-1. A range of values is given in some instances, because the observed radius depends upon the way in which the atom is covalently bonded. For example, the van der Waals radius of an H atom varies from 1.0 Å when bonded to an aromatic C atom, to 1.54 Å when bonded to a negative ion. Fortunately, these are extreme variations, and **two atoms are generally in close van der Waals contact when the distance between their centers is approximately 0.8 Å greater than when they are covalently bonded.** The van der Waals radius is a minimum estimate of the size of the atom, and **optimal van der Waals attractions to other molecules** (Section 2.3) **generally occur at a radius that is about 0.2 Å greater than the van der Waals radius.**

Table 2-1. Values of the van der Waals radius (r_{vdw}) and covalent radius (r_{cov}) of common elements in biochemistry

Element	r_{vdw} **(Å)**	r_{cov} **(Å)**
H	1.0–1.54	0.37
He	1.50–1.80	0.30
Li	1.80	1.34
C	1.65–1.70	0.77
N	1.55	0.75
O	1.50	0.73
F	1.35–1.60	0.71
Na	2.30	1.54
Mg	1.70	1.45
P	1.85–1.90	1.10
S	1.80	1.20
Cl	1.70–1.90	0.99
K	2.80	1.96
I	1.95–2.15	
Organic groups		
CH_3	2.00	
C_6H_6	1.70 (perpendicular to the ring)	

Data are from J. Israelachvili (1991) *Intermolecular and Surface Forces*, 2nd edn, Academic Press, London, pp. 176–212; J. E. Huheey (1983) *Inorganic Chemistry: principles of structure and reactivity*, 3rd edn, Harper and Row, NY, Chapter 6; G. M. Barrow (1988) *Physical Chemistry*, 5th edn, McGraw-Hill, NY, pp. 30–53; A. Bondi (1964) *J. Phys. Chem.* **68**, 441–451.

The size of molecules. A. Y. Meyer (1986) *Chem. Soc. Rev.* **15**, 449–474.

2.1.A. Molecular Surfaces and Volumes

With larger molecules composed of many atoms, each atom is depicted as a sphere of the appropriate van der Waals radius; where the atoms are covalently bonded, overlapping regions of the spheres are truncated. The complex surface that results is called the **van der Waals surface**. It has a strictly defined surface area and encloses a definite volume; some pertinent values for various atoms and chemical groups are given in Table 2-2. The volumes and surface areas of entire molecules may usually be estimated by summing the parameters for the constituent parts, so long as the molecule is not structurally strained.

Table 2-2. van der Waals surface areas and volumes of chemical groups when bonded to C atoms

Chemical group	Surface area (Å²)	Volume (Å³)
>C<	1.0	5.5
>CH–	10.9	11.5
$-CH_2-$	20.9	16.8
$-CH_3$	33.4	22.3
$-C_6H_5$	94.9	76.1
-OH	19.3	12.6
–C(=O)–	22.3	18.2
–C(=O)–OH	43.4	
-SH		24.6
-NH-	26.5	17.5
$-NH_2$	16.4	13.4

From T. E. Creighton (1993) *Proteins: structures and molecular properties*, W. H. Freeman, NY.

The van der Waals surface is not very relevant to a folded macromolecule like a protein or nucleic acid, where internal atoms and nooks and cavities between the atoms are normally not accessible to the solvent or any other molecules present. No chemical procedure can measure the van der Waals area or volume directly, because any chemical probe has significant dimensions. **A more relevant surface is that which normally is in contact with the solvent.** This can be defined by rolling a spherical probe, representing a solvent molecule, over the van der Waals surface of the macromolecule (Figure

2-1). Those parts of the van der Waals surface in contact with the surface of the solvent molecule are designated the **contact surface**. When the probe is simultaneously in contact with more than one atom of the macromolecule, its interior surface defines the **reentrant surface**. The contact surface and the reentrant surface together make a continuous surface, which is defined as the **molecular surface**, and it defines a **molecular volume**. The surface defined by the center of the probe molecule is the **accessible surface**. In Figure 2-1 the probe does not contact atoms 3, 9 or 11, and they have no accessible surface area. Such atoms are considered to be interior atoms, not part of the surface of the molecule.

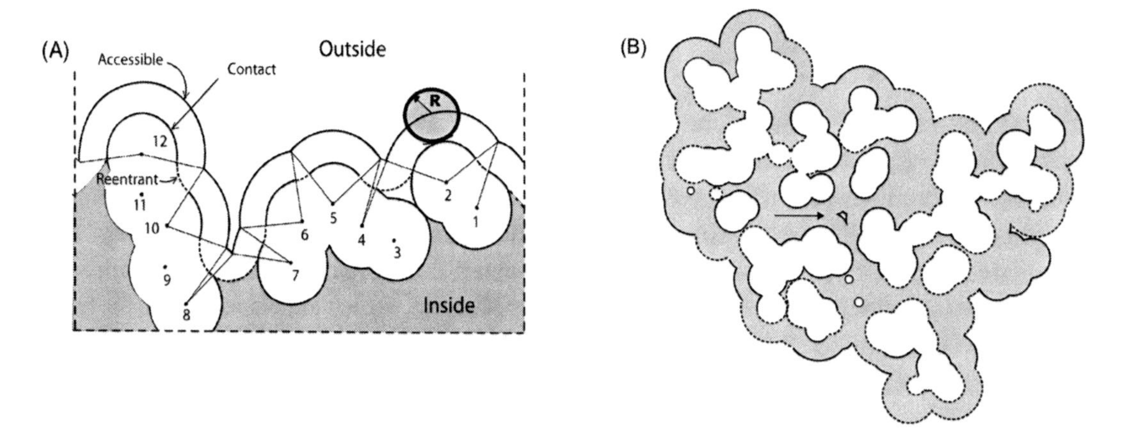

Figure 2-1. Analysis of molecular surfaces. (A). Definition of the various surfaces for a two-dimensional slice through atoms 1–12 of a hypothetical macromolecule. The sphere of radius R is the probe that is rolled over the surface in van der Waals contact with all possible atoms of the macromolecule. (B). The van der Waals and accessible surfaces of a section through a folded protein. The outer surface is the accessible surface, the inner surface is the van der Waals surfaces of the individual atoms. The solid lines outline C and S atoms; the dashed lines N and O. In places, the accessible surface is controlled by atoms above or below the section shown. The arrow indicates a cavity inside the molecule large enough to accommodate a solvent molecule with a radius of 1.4 Å, which was used to define the accessible surface. Adapted from F. M. Richards.

The accessible surface area depends on the size of the probe. If the radius of the probe in Figure 2-1 were to increase about three-fold, the number of atoms that contact the probe would decrease from about 9 to 3. In addition, the accessible surface would be much smoother. Thus, the smaller the probe, the larger the accessible surface area, the smaller the volume and the rougher the surface. Accessible surface areas are usually measured with a probe with a radius of 1.4 Å, representing a water molecule.

Properties of atoms in molecules: atomic volumes. R. F. W. Bader *et al.* (1987) *J. Am. Chem. Soc.* **109**, 7968–7979.

2.1.B. Packing Density

The van der Waals radii of atoms also permit calculation of the packing density of a substance or a macromolecule. **The packing density is the fraction of space that is occupied by the atoms, as defined by their van der Waals radii.** For reference, close packing of identical spheres produces a packing density of 74%, and crystals of small molecules typically have values of 70–80%. Liquids generally have lower packing densities, such as 44% for the nonspherical cyclohexane.

The packing density in proteins: standard radii and volumes. J. Tsai *et al.* (1999) *J. Mol. Biol.* **290**, 253–266.

An estimate of random close packing density in monodisperse hard spheres. P. Jalali & M. Li (2004) *J. Chem. Phys.* **120**, 1138–1139.

2.2. ELECTROSTATIC FORCES: SIMPLICITY TO COMPLEXITY

The most fundamental noncovalent attraction between atoms and molecules is that between electrostatic charges, and **all intermolecular forces are believed to be essentially electrostatic in origin**. Positive and negative charges attract, whereas the same charges repulse. The unit of charge corresponding to a single electron or proton is designated as $-e$ or $+e$. Molecules or groups of atoms that contain significant positive or negative charges, even if the charges are only partial, are said to be **polar**. These molecules or groups often contain O, S or N atoms, which are **electronegative** (retaining electrons) and nucleophiles; they tend to induce **polarization** of the electrons within the molecule and generate partial charges. Polar groups can react chemically with other such groups or act as acceptors or donors of hydrogen bonds (Section 2.4). Molecules or groups that contain full electrostatic charges, such as the $-NH_3^+$ and $-COO^-$ groups of amino acids in neutral solution, are especially polar.

The opposite character is to be **nonpolar**, lacking electronegative atoms like N, O and S and not containing groups with substantial partial electrostatic charges. Such molecules typically are hydrocarbons, containing only C and H atoms. Generally they are unreactive chemically.

Molecular dynamics simulations of biomolecules: long-range electrostatic effects. C. Sagui & T. A. Darden (1999). *Ann. Rev. Biophys. Biomol. Struct.* **28**, 155–179.

Electrostatic aspects of protein–protein interactions. F. B. Sheinerman *et al.* (2000) *Curr. Opin. Struct. Biol.* **10**, 153–159.

Close-range electrostatic interactions in proteins. S. Kumar & R. Nussinov (2002) *Chembiochem.* **3**, 604–617.

2.2.A. Point Charges: the Simplest Interaction

The energy of the electrostatic interaction between two atoms, A and B, is given by **Coulomb's law** and is simply the product of their two charges divided by the distance between them, r_{AB}:

$$\Delta E = \frac{Z_A Z_B e^2}{r_{AB}} \tag{2.1}$$

e is the charge of an electron and Z is the number and type of such charges on each atom. The energy, ΔE, is expressed relative to that when the two charges are very far apart.

If the two charges are of opposite sign, the energy decreases as they approach each other: the interaction is favorable and they attract each other. If the charges are of the same sign, they repulse each other. **The electrostatic interaction is effective over relatively large distances because it varies inversely with only the first power of the distance**; at twice the distance, the interaction energy is only halved. It is also a very strong interaction; two point charges 5.5 Å apart in a vacuum would have an interaction of 1.0×10^{-19} calories, which is equivalent to 60 kcal/mol (256 kjoules/mol).

Coulomb's law as stated in Equation 2.1 is valid only in a vacuum. In other environments, the electrostatic interaction is modulated by other interactions with the environment. In homogeneous environments, this can be expressed by the **permittivity** or the **dielectric constant**, D:

$$\Delta E = \frac{Z_A Z_B e^2}{D r_{AB}} \qquad (2.2)$$

The values of **dielectric constants are invariably greater than unity, so the electrostatic interaction is always diminished in media other than a vacuum**. This is especially important with liquids, which usually have dielectric constants in the range of 2–110. The dielectric constant reflects the degree of **polarizability** of the material (Section 2.2.B.2). A polar liquid has a much greater dielectric constant than a nonpolar liquid. For example, the dielectric constants at 20° C of water and cyclohexane are 80 and 2, respectively.

The concept of a dielectric constant is valid only for homogeneous environments; less homogeneous environments need to be treated explicitly. Special problems arise at the interface between regions with very different bulk dielectric constants. For example, two charges on opposite sides of a sphere of low dielectric constant immersed in a medium of high dielectric constant, such as water, interact not through the shortest distance, through the low dielectric, but around the outside of the sphere. The very long path length through the high dielectric constant results in the energy of this electrostatic interaction being much smaller than might otherwise be expected.

Coulomb's law (Equation 2.1) ignores the finite sizes of ions, so it holds only at distances significantly greater than atomic dimensions. The charge of an atom is separated between the nucleus and the diffuse electron cloud, so it cannot be treated as a point charge at short distances. This problem is even more severe with organic molecules in which the charge is frequently distributed over two or more H, N or O atoms. Electrostatic interactions become especially complex in aqueous solutions containing salts, the usual milieu of molecular biology, where individual ions interact with the charged groups (Section 3.5).

2.2.B. Dipoles: Charge Separation Within a Molecule

A molecule need not have a net charge to participate in electrostatic interactions, because **the electron density can be localized nonuniformly if the atoms that are covalently bonded have different electronegativities**. Of two atoms linked by a covalent bond, that with the greater electronegativity has an excess of negative charge, δ^-, and the other atom has an excess of positive charge, δ^+, generating a **dipole**:

$$\text{O–H} \leftrightarrow \text{O}^{\delta-}\text{–H}^{\delta+} \qquad (2.3)$$

The relative **electronegativities** of the atoms in biological macromolecules are:

O	3.45	(2.4)
N	2.98	
C	2.55	
S	2.53	
P	2.2	
H	2.13	

The importance of electronegativities is illustrated by the peptide bond of proteins, which has a partial double-bond character due to resonance with a form in which the more electronegative O atom acquires a negative charge, while the –NH– group is positively charged:

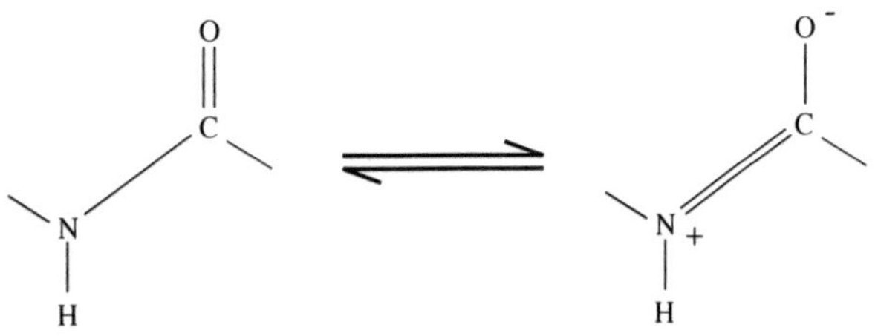

$$-\mathrm{NH}-\mathrm{C}(=\mathrm{O})- \rightleftharpoons -\mathrm{N}^{+}\mathrm{H}=\mathrm{C}(-\mathrm{O}^{-})- \qquad (2.5)$$

The double-bonded, charged species is populated about 40% of the time, so the peptide group can be represented as having partial charges of as much as ±0.4 *e*. Polar O and N atoms in other molecules have partial charges as great as ±0.35 *e*, but those of atoms in aliphatic molecules are probably no greater than ±0.1 *e*.

The π electrons in aromatic rings, such as those of benzene, are localized above and below the face of the ring. **This excess of electrons gives the face of the aromatic ring a small net negative charge** of approximately –0.15 *e*, while the H atoms on its edge have a corresponding positive net charge. The electrostatic interactions between these partial charges dominate the interactions between aromatic rings. Such rings prefer to interact with the positively charged edge of one ring pointing at the negatively charged face of another, or with their rings parallel but offset, so that the edge of each ring is interacting with the face of the other. Aromatic rings do not interact as favorably by stacking their rings one above the other using nonpolar interactions. Electronegative O and S atoms tend to interact favorably with the slight positive charge at the edges of aromatic rings, whereas –NH– groups tend to interact with the π electrons at their faces.

Attractive intramolecular edge-to-face aromatic interactions in flexible organic molecules. W. B. Jennings *et al.* (2001) *Acc. Chem. Res.* **34**, 885–894.

A group electronegativity equalization scheme including external potential effects. T. Leyssens *et al.* (2006) *J. Phys. Chem.* **110**, 8872–8879.

Estimation of electronegativity values of elements in different valence states. K. Li & D. Xue (2006) *J. Phys. Chem.* **110**, 11332–11337.

1. Dipole Moment

The separation of charge within a molecule determines its **dipole moment,** μ_D. Its magnitude is given by the product of the separated excess charge, Z, and the distance, *d*, by which it is separated:

$$\mu_D = Zd \tag{2.6}$$

One electron equivalent of positive and negative charge separated by 1 Å has a dipole moment of 4.8 Debye units (D). The peptide bond (Equation 2.5) has a dipole moment of 3.5 D, that of a water molecule 1.85 D. **The dipole moment has directionality as well as magnitude and is usually depicted as a vector along a straight line from the negative to the positive charge.** The dipole moment of the peptide bond can be represented as:

where the dashed bonds indicate the resonance forms of Equation 2.5.

Dipoles interact with point charges, with other dipoles and with more complex charge separations known as **quadrupoles, octupoles,** etc., in a complex manner that depends upon the relative orientations of the various groups. The interactions can be computed by considering the individual charges, including those of the dipole, quadrupole, etc., and calculating the Coulombic interactions between all of them according to Equation 2.2. The interactions between the four partial charges of two dipoles are analogous to those between two bar magnets. Two dipoles side-by-side will repel each other when parallel, whereas there will be an equivalent attraction between them when antiparallel. Maximum interactions occur in a head-to-tail orientation, being either repulsive or attractive. **Dipolar interactions are weaker than those between ions,** because both attractions and repulsions occur with the two separated charges of each of the dipoles. This also has the effect of making the energy of the interactions depend inversely upon the second to third power of the distance between the interacting molecules when in fixed orientations, and upon the sixth power when the molecules are free to rotate in response to the interaction. As a result, electrostatic interactions involving dipoles decrease much more abruptly as the distance is increased than do interactions between point charges. The interactions are, however, attenuated in the same way by the dielectric properties of the medium (Equation 2.2).

The multiple interactions that take place between point charges and dipoles on a number of atoms are mutually dependent and turn a very simple relationship like Coulomb's law (Equation 2.1) into a very complex phenomenon. The electrostatic interactions between molecules in a homogeneous liquid can be averaged and expressed as a simple dielectric constant of the liquid, but not when the environment is not homogeneous at the molecular level, which is always the case with the complex structures of proteins and nucleic acids immersed in water. In these cases, electrostatic interactions invariably involve interactions between multiple charges and dipoles of the macromolecule, and between these and the bulk solvent and any ions in it; the net effects usually depend upon a delicate balance between charge–charge and charge–solvent interactions. In this case, interactions between individual charges and dipoles of the protein must be calculated directly.

2. *Polarizability*

Electrostatic interactions involving dipoles modify the nature of the charge distribution of the dipole in the interacting molecules, which is simply an unequal distribution of electrons and is easily perturbed. **An electric dipole is induced in even a neutral, nonpolar molecule or group by an external electric field E**, because the external field modifies the electrostatic balance between the electrons and the nucleus in each atom. The magnitude of this induced dipole is given by:

$$\mu^{ind} = \alpha_0 E \tag{2.8}$$

where the coefficient α_0 is the electronic **polarizability**. The value of α_0 for any particular molecule depends primarily on how tightly the electrons are held by the nuclei of the atoms. In general, **the larger an atom, the greater its polarizability**, because the electrons are held less tightly by the nucleus. An important aspect of polarizability is that **the induced dipole always interacts favorably with the field that induced it, so the result is an attraction between them** (Section 2.3). The energy of this interaction is only half what would have occurred if the dipole pre-existed, because some of the energy of interaction must be used in inducing the dipole.

When the molecule or group has additional permanent dipoles, they also reorient due to the external field. Consequently, the polarizability of a molecule is the sum of the electronic and orientational polarizabilities. The contribution from the orientational polarization of a polar group is generally much larger than that of the electronic polarization.

On the role of polarizability in chemical–biological interactions. C. Hansch *et al.* (2003) *J. Chem. Inf. Comp. Sci.* **43**, 120–125.

Effective molecular polarizabilities and crystal refractive indices estimated from X-ray diffraction data. A. E. Whitten *et al.* (2006) *J. Chem. Phys.* **125**, 174505.

2.2.C. Ion Pairs and Salt Bridges

An **ion pair** is two molecules or groups of opposite electric charge that are in close proximity due to the favorable electrostatic interactions between them. A **salt bridge** is an ion pair that contact each other at their closest possible approach, at the sum of their van der Waals radii. Very close interactions between oppositely charged groups in macromolecules usually consist not only of electrostatic interactions but also of at least some element of hydrogen bonding (Figure 2-2). This type of interaction is known as a **salt linkage** or an **ionic bond**.

Relationship between ion pair geometries and electrostatic strengths in proteins. S. Kumar & R. Nussinov (2002) *Biophys. J.* **83**, 1595–1612.

Protein stabilization by salt bridges: concepts, experimental approaches and clarification of some misunderstandings. H. R. Bosshard *et al.* (2004) *J. Mol. Recognit.* **17**, 1–16.

Statistical characterization of salt bridges in proteins. J. N. Sarakatsannis & Y. Duan (2005) *Proteins* **60**, 732–739.

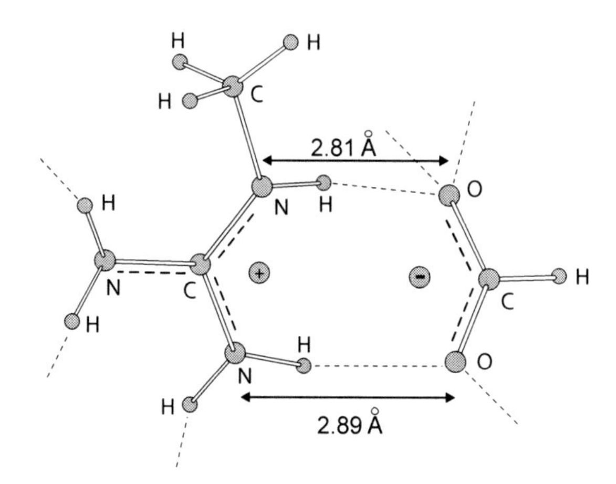

Figure 2-2. Hydrogen bonds involved in the interaction between ionized carboxyl and guanidinium groups, illustrated by the crystal structure of methylguanidinium formate. The thin dashed lines indicate hydrogen bonds assumed to be present from the short distances between the adjacent atoms. The thick dashed lines indicate partial double bonds due to resonance in the molecule. Data from D. D. Bray.

2.3. VAN DER WAALS INTERACTIONS: THE ADVANTAGES OF CLOSE PACKING

There is an attractive force between all atoms and molecules, even in the absence of charged groups, as a result of mutual interactions related to the induced polarization effects described in Section 2.2.B.2. These ubiquitous attractions are known as **van der Waals interactions**. They are weak and close-range, varying as the inverse sixth power of the distance between the centers of the interacting atoms, d^6. They arise from three types of interactions: (1) between two permanent dipoles, (2) between a permanent dipole and an induced one, and (3) between two mutually induced dipoles, known as **London dispersion forces**. The first two have been described above, but the third is the more important in that it occurs between all atoms and molecules.

The London dispersion force is quantum mechanical in nature, but a greatly simplified description can be derived from the classical picture of an atom with electrons orbiting the nucleus. A spherical atom has no net dipole moment, but it can have a transient instantaneous dipole resulting from a temporary asymmetric orientation of the electrons and nucleus. This transient dipole polarizes any neutral atom nearby, creating an attraction between them. Although the transient dipole of the first atom is constantly and rapidly changing, that of the other atom tends to follow it, and the two are correlated. This dispersion force is basically electrostatic in nature and depends upon d^{-6}, as do the other two components of the van der Waals interactions described above. This distance dependence breaks down at distances greater than about 50 Å, becoming proportional to d^{-7}, for the correlation between the two electron distributions diminishes because of the time it takes the electrostatic field from one atom to reach the other; at such distances, the interaction is very weak. The van der Waals interactions between objects larger than atoms have less pronounced dependence upon the distance between them; for example, the interaction between two large spheres is proportional to the simple inverse of the distance between them (Figure 2-3).

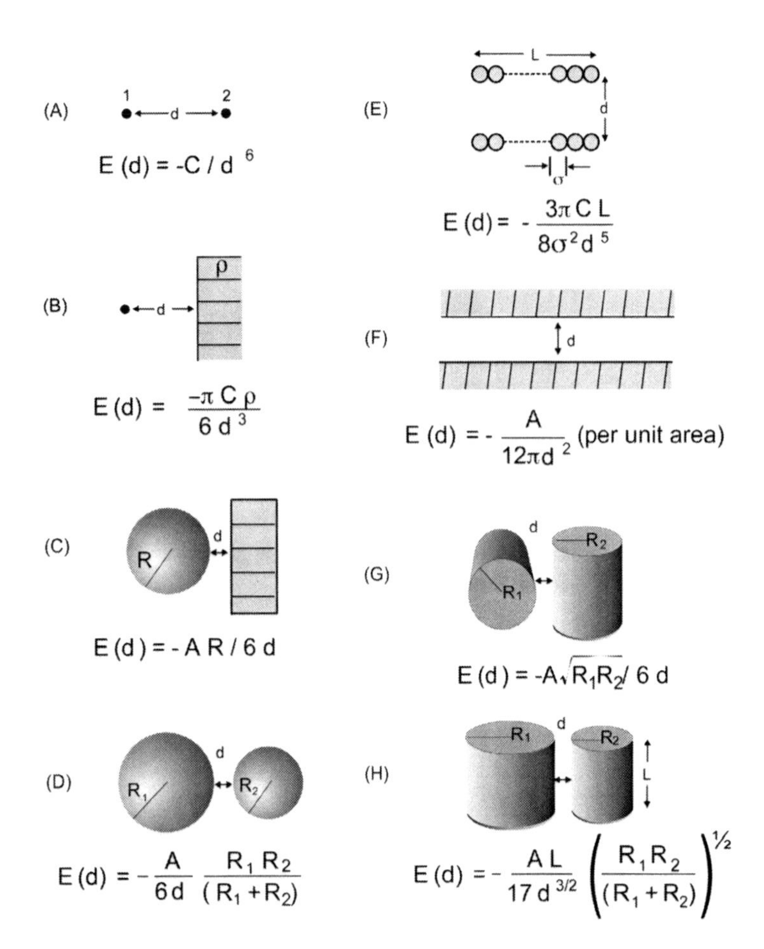

Figure 2-3. Dependence of van der Waals potential, E(d), between two objects with finite sizes or with infinite extensions. d is the distance between the two objects; A and C are constants; ρ is the density of the object. Adapted from J. Israelachvili.

The van der Waals interactions between atoms are often represented by an energy potential as a function of distance between them, E(d), that includes both the attractive force and the repulsion at close range (Section 2.1). The most well-known of these is the **Lennard–Jones potential** of the form:

$$E(d) = \frac{C_n}{d^n} - \frac{C_6}{d^6} (n > 6) \tag{2.9}$$

where C_n and C_6 are constants. The first term gives the repulsions, the second the van der Waals attractions. The most common potential has $n = 12$, which is efficient in computations and known as the **Lennard–Jones 6,12 potential** (Figure 2-4).

The distance for the optimal interaction of two atoms is given by the minimum in Figure 2-4. It is usually 0.3–0.5 Å greater than the sum of their van der Waals radii measured from their closest contact distance in crystals (Table 2-1). The van der Waals radius is defined by the steeply ascending repulsive interaction at closer distances. The van der Waals interaction is generally considered to be independent of the orientation of the interacting molecules, but this is only approximately true, especially when the interacting molecules are independent and tumbling rapidly in a gas or a liquid. The magnitude of the interaction of even a nonpolar group like $-CH_3$ can vary with orientation, because the polarizability of a C–H bond is nearly twice as great along the bond as perpendicular to it. With nonspherical molecules, the variation of the van der Waals interaction with distance and orientation is much more complex (Figure 2-3).

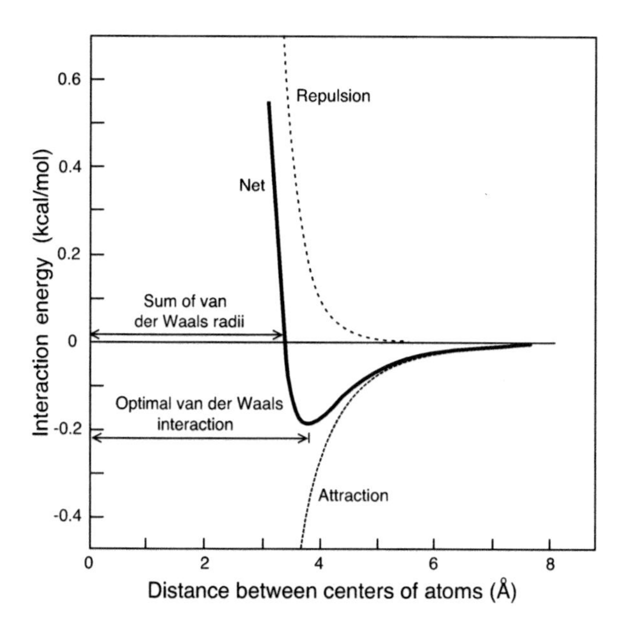

Figure 2-4. Representative profile of the energy of the van der Waals interaction as a function of the distance, *r*, between the centers of the two atoms. The individual attractive and repulsive components are indicated by the dashed lines, the net interaction by the solid line. The optimal interaction between the two atoms occurs where the energy is at a minimum. The sum of the van der Waals radii of the two atoms is given by the distance where the energy increases sharply. The interaction energy was calculated using the Lennard–Jones 6,12 potential (Equation 2.9) with C_{12} = 2.75 × 10^6 $Å^{12}$ kcal/mol and C_6 = 1425 $Å^6$ kcal/mol, values considered appropriate for the interaction between two C atoms. From T. E. Creighton (1993) *Proteins: structures and molecular properties* 2nd edn, W. H. Freeman, NY, p. 146.

Analytical dispersion force calculations for nontraditional geometries. S. W. Montgomery *et al.* (2000) *J. Colloid Interface Sci.* **227**, 567–584.

Van der Waals interactions in a dielectric with continuously varying dielectric function. R. Podgornik & V. A. Parsegian (2004) *J. Chem. Phys.* **121**, 7467–7473.

Van der Waals dispersion forces between dielectric nanoclusters. H. Y. Kim *et al.* (2007) *Langmuir* **23**, 1735–1740.

2.4. HYDROGEN BONDS: SPECIFICITY AND DIRECTIONALITY

A hydrogen bond occurs when two electronegative atoms compete for the same H atom:

$$\text{–D–H} \cdots \text{A–} \tag{2.10}$$

The dotted line is the hydrogen bond. The H atom is formally bonded covalently to one of the atoms, the **donor** D, but it also interacts favorably with the other, the **acceptor** A. Some examples are illustrated in Figure 2-5. In a few strong, short hydrogen bonds, the H atom is symmetrically placed between the two electronegative atoms, but normally the H atom remains covalently bonded and closer to one of the atoms, with a normal covalent bond length.

The main component of the hydrogen bond is an electrostatic interaction between the dipole of the covalent bond to the H atom, in which the H atom has a partial positive charge, and a partial negative charge on the other electronegative atom:

$$–\ D^{\delta-}–\ H^{\delta+} \cdots {}^{\delta-}A\ – \tag{2.11}$$

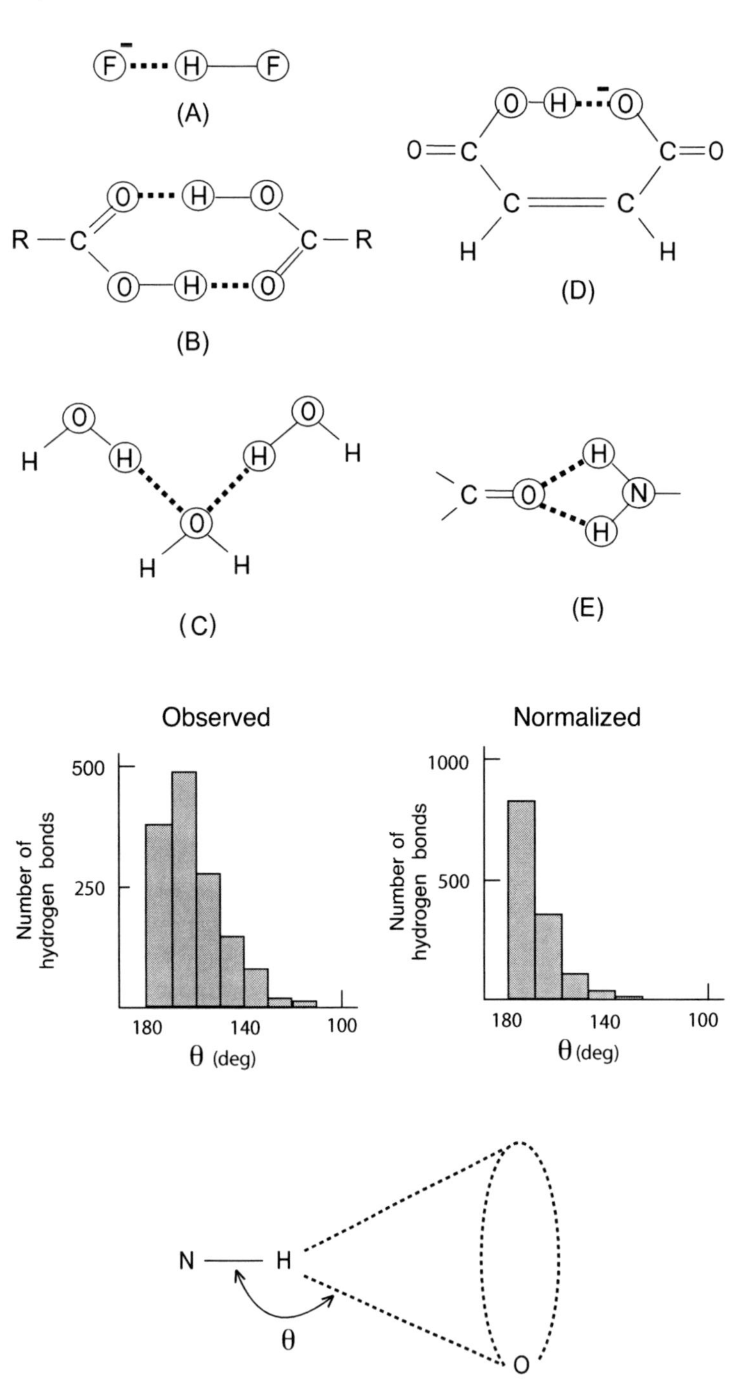

Figure 2-5. Various geometries of hydrogen bonding interactions. The hydrogen bonds are the dashed lines; atoms involved in the hydrogen bond are circled. (A) A hydrogen bond between a fluoride anion and hydrogen fluoride. (B) A cyclic dimer of carboxyl groups; note the symmetry. (C) Hydrogen bonding between water molecules; all the water molecules can participate in such hydrogen bonding, as in ice (Figure 3-3) and, to a lesser extent, liquid water. (D) Intramolecular hydrogen bonding; note that the charge and H atom can alternate between the two O atoms. (E) The single O atom of a carbonyl group is bonded simultaneously to both H atoms of an amino group; such hydrogen bonds are common in biological macromolecules.

Figure 2-6. Linearity of N–H···O hydrogen bonds observed in crystal structures of small molecules. The degree of linearity is measured by the angle θ (*bottom*), which would have a value of 180° for a perfectly linear hydrogen bond. The histogram of observed values of intermolecular hydrogen bonds (*left*) is affected by a geometric factor in which the various ranges of the value of θ include different volumes of three-dimensional space. Correcting for this gives the histogram on the *right*, which illustrates the marked tendency of hydrogen bonds to be linear. Data from R. Taylor & O. Kennard.

The H atom is special in being able to interact strongly with electronegative atoms, while still being covalently attached to another. This is due to its small size but substantial charge, resulting from its tendency to be positively polarized. An additional aspect of the hydrogen bond arises from a transfer of electrons, as in a covalent bond, and a **typical hydrogen bond is believed to have roughly 10% covalent nature.** Both the electrostatic and covalent aspects cause the most commonly observed, and presumably most energetically favorable, hydrogen bonds to have the three atoms colinear (Figure 2-6). There is considerable uncertainty, however, about how the strength of the hydrogen bond interaction varies with departures from linearity.

The chemical groups in biomolecules that most commonly serve as hydrogen bond donors are N–H, O–H and, much less frequently, S–H and C–H. The most common acceptors are O=, –O–, –N= and, much less frequently, $-S^-$, –S– and the π electrons of aromatic groups. O atoms are frequently observed to participate simultaneously as the acceptor in two hydrogen bonds, as in Figure 2-5-C and -E. Much less frequently, a single H atom from a donor can be shared between two acceptors. The partial negative charge at the electronegative acceptor atom, such as an O atom, is localized on the lone-pair electron orbitals, and an intrinsic preference for the H atom to point toward these electrons might be expected. Hydrogen bonds between two molecules in the gas phase show this geometry, and it is also frequently observed in crystals with carbonyl oxygen acceptors (Figure 2-7). In other cases, however, it is not observed so frequently, and the energetic preference for H atoms to be directed at the lone-pair electrons is probably so small as to be easily overwhelmed by other considerations.

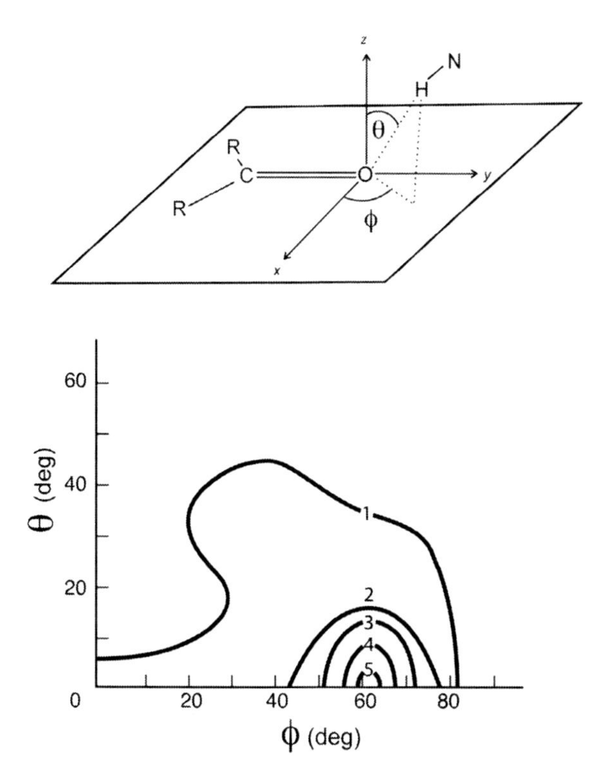

Figure 2-7. The geometries of –C=O···H–N– hydrogen bonds observed in crystal structures of small molecules. The definitions of the angles ϕ and θ are illustrated at the *top*, and the relative frequencies of their observed values in intermolecular hydrogen bonds are given by the contours. The angle ϕ measures departures from linearity of the C=O bond and the H atom; the most frequently observed values are in the region of 50°–60°. The angle θ measures the extent to which the H atom lies out of the plane defined by the R, C and O atoms; the most commonly observed values are in the region of 0°–7°. The lone-pair electrons of the O atom are believed to project at angles of ϕ = 60°, θ = 0°. The spherical polar coordinate system used here gives a bias towards small of values of θ that could be corrected by plotting sin θ. Data from R. Taylor & O. Kennard.

The lengths of hydrogen bonds, and their strengths, depend upon the electronegativities of the acceptor and donor; the greater their electronegativities, the shorter the distance between them and the stronger the hydrogen bond. Charged groups also give shorter and stronger hydrogen bonds. The optimal lengths of various types of hydrogen bonds are given in Table 2-3. In strong, short hydrogen bonds, known as 'low-barrier' hydrogen bonds, the H atom is located midway between the donor and acceptor atoms, which are separated by no more than 2.3–2.5 Å.

The strengths of hydrogen bonds are generally said to be within the rather broad range of 2–10 kcal/mol (8–40 kJ/mol) at room temperature. Part of this variation is due to the variety of hydrogen bonds, but much is also a result of uncertainty because there is no direct way to measure their strengths.

Table 2-3. Typical dimensions of hydrogen bonds

Hydrogen bond	Distance between atoms (Å)		
X – H••••Y	X – Y	X – H	H – Y
O – H••••O	2.75	1.00	1.7
N – H••••O	2.99	0.90	2.1
O – H••••N	2.78	1.10	1.7
N – H••••N	3.08	0.90	2.2
P – OH••••O = P			1.55–1.69
P – OH••••OH			1.65–1.89
O_w – H••••O = P			1.66–1.88
P – O – H••••O_w			1.59–1.68
Other O – H••••O			1.74–2.18
N – H••••O = P			1.58–1.89
N – H••••O = C			1.69–2.32
N – H••••N			1.73–2.23
NH – H••••O = P			1.67–2.07
NH – H••••O = C			1.68–2.76
NH – H••••N			1.85–2.76

O_w = oxygen atom of water molecule.

The ionic hydrogen bond. M. Meot-Ner (2005) *Chem. Rev.* **105**, 213–284.

Potential functions for hydrogen bonds in protein structure prediction and design. A. V. Morozov & T. Kortemme (2005) *Adv. Protein Chem.* **72**, 1–38.

Estimates of the energy of intramolecular hydrogen bonds. M. Jablonski *et al.* (2006) *J. Phys. Chem. A* **110**, 10890–10898.

Geometry, energetics, and dynamics of hydrogen bonds in proteins: structural information derived from NMR scalar couplings. J. Gsponer *et al.* (2006) *J. Am. Chem. Soc.* **128**, 15127–15135.

Watching hydrogen-bond dynamics in a β-turn by transient two-dimensional infrared spectroscopy. C. Kolano *et al.* (2006) *Nature* **444**, 469–472.

2.5. INTRAMOLECULAR INTERACTIONS: THE IMPORTANCE OF ENTROPY

For two or more independent molecules to interact, they must become fixed in space relative to each other. In other words, they must lose entropy (Section 1.4), which is energetically unfavorable. Were it not for entropy, all matter would be solid. The entropic contribution to the free energy arises from molecules having freedom, and it makes possible the liquid and gas states. How much entropy must be lost in an interaction depends upon the number of degrees of freedom that must be fixed. For example, van der Waals interactions require the least entropy loss, since only the distance between two atoms needs to be fixed (Figure 2-4), whereas hydrogen bonding requires that both proximity and orientation be fixed to some extent (Figures 2-2, 2-5, 2-6 and 2-7).

Entropic contributions are very important when two or more interactions can occur simultaneously, **because in favorable cases much less entropy need be lost in forming the second and subsequent interactions than in forming the first**. Two interactions that can occur simultaneously can be much more favorable energetically than might be expected from their individual strengths.

To calculate the entropies of macromolecules is unfortunately not yet practical. Consequently, the description that will be presented here is simple and empirical.

Entropy in protein folding and in protein–protein interactions. G. P. Brady & K. A. Sharp. (1997) *Curr. Opinion Struct. Biol.* **7**, 215–221.

Intramolecular interactions at protein surfaces and their impact on protein function. A. D. Robertson (2002) *Trends Biochem. Sci.* **27**, 521–526.

2.5.A. Effective Concentrations: an Empirical Approach to the Entropy Problem

The magnitude of **entropic cooperativity can be illustrated with intramolecular interactions**. Two parts of the same molecule can interact without all the loss of entropy that is required to bring two independent molecules together; the two parts of a single molecule are already fixed in proximity and orientation to varying degrees by the covalent bonds. Depending upon the molecule and the interaction, only a fraction of the internal flexibility of the molecule will need to be lost for the interaction to occur intramolecularly.

Interactions between model compounds dissolved in aqueous solution are measured by the equilibrium constant, K_{AB}, for the association of two appropriate individual molecules, A and B, to form the complex A•B:

$$A + B \xleftrightarrow{K_{AB}} A \bullet B \tag{2.12}$$

$$K_{AB} = \frac{[A \bullet B]}{[A][B]} \tag{2.13}$$

The strengths of intramolecular and bimolecular examples of the same interaction can be compared quantitatively by the ratio of their equilibrium constants:

$$A\text{—}B \xrightleftharpoons{K_{intra}} A \bullet B \text{ (cyclic)} \tag{2.14}$$

$$A + B \xrightleftharpoons{K_{inter}} A \bullet B \tag{2.15}$$

That for the intramolecular case is dimensionless, while that for the intermolecular interaction has dimensions of (concentration)$^{-1}$. Therefore, the ratio of the two has the dimensions of concentration, and it can be thought of as the **effective concentration** (or **effective molarity**) of the two groups relative to each other when part of the same molecule:

$$\frac{K_{intra}}{K_{inter}} = \text{effective concentration of A—B} \quad (2.16)$$

It was long thought that the maximum effective concentration in aqueous solution was about 55 M, the concentration of pure water, when one group could be considered to be immersed in a liquid environment of the second component. Consequently, and unfortunately, one often finds instances of intra- and intermolecular interactions being interconverted using this factor of 55 M.

Many experimental measurements have been made for various inter- and intramolecular chemical reactions, and very large values of effective concentrations are generally measured; a few representative examples are given in Table 2-4. They represent chemical reactions involving reversible covalent bond formation that can be considered analogous to noncovalent interactions. On the other hand, the covalent nature of these interactions probably exaggerates the entropic effect, due to the more stringent geometrical requirements for covalent bond formation than for other types of interactions. In any case, the first three examples (A–C) of Table 2-4 involve flexible molecules with relatively free rotations about three single bonds that must be restricted to form the product. In spite of this considerable flexibility in the reactant, and the entropic loss in the reaction, the effective concentrations measured are in the range of 10^{+3}–10^{+5} M. Therefore, **merely keeping two groups in reasonable proximity by linking them covalently through several covalent bonds causes their concentration relative to each other to be much higher than would be practical with them on separate molecules, even in the most concentrated liquid state.** The last example (D of Table 2-4) has an enormous effective concentration of $5 \times 10^{+9}$ M. This undoubtedly is due primarily to the small difference in entropy between the molecule with and without the anhydride interaction. In this case, the planar aromatic structure of the molecule keeps the carboxyl groups in close proximity, whether or not the anhydride is present. The very small loss in flexibility and entropy that occurs in the reaction results in an enormous effective concentration that is close to the maximum considered possible theoretically (approximately 10^{10} M).

When there is no entropic difference between the molecules with and without the interaction, the effective concentration has its maximum value. This value depends upon the type of interaction. Those where the proximity and orientation of the interacting groups are very important, as in a hydrogen bond, and especially when a covalent bond is formed, have very high maximum effective concentrations. Where these factors are not so important, as in van der Waals interactions, the groups still have significant degrees of freedom when interacting and have less entropy to gain upon dissociating, so lower values of maximum effective concentrations apply. Even in this case, though, the maximum values are substantially greater than 55 M. The reason for this is that the molecules of a liquid have a high degree of both rotational and translational freedoms, so they are usually not in the optimal orientation for interacting with each other.

The magnitudes of the effective concentrations expected for interactions of the type observed in proteins and nucleic acids are unfortunately not known. Only in the case of the disulfide interaction between thiol groups have values been measured in proteins. **The maximum value measured with protein disulfide bonds is somewhat greater than 10^5 M**, but the disulfide bond is a type of covalent

Table 2-4. Selected examples of measured values of effective concentrations of two reactive groups in small molecules

Example	Equilibrium reaction	Effective concentration (M)
(A)	CO_2H–CH_2–CH_2–CH_2–SH ⇌ H_2C–C(H_2)–CH_2–C(=O)–S– (ring) + H_2O	3.7 X10^3
(B)	SH–CH_2–HOCH–HOCH–CH_2–SH + GSSG ⇌ HOCH–HOCH–C(H_2)–S–S–C(H_2)– (ring) + 2GSH	2 X10^2
(C)	CO_2H–CH_2–CH_2–CO_2H ⇌ H_2C–H_2C–C(=O)–O–C(=O)– (ring) + H_2O	1.9 X10^5
(D)	(CH_3, CO_2H, CO_2H, CH_3 substituted benzene) ⇌ (benzene fused C(=O)–O–C(=O) ring) + H_2O	5.4 X10^9

From T. E. Creighton (1993) *Proteins: structures and molecular properties*, 2nd edn, W. H. Freeman, NY.

interaction and they tend to exhibit high maximum effective concentrations. Hydrogen bonds are moderately sensitive to orientation and probably have a partial covalent character, so substantial values would be expected, but probably much less than those of 10^{10} M and less than those involving disulfide bonds. Ionic and hydrophobic interactions are not very stringent stereochemically, so maximum values of 10^2–10^3 M may apply in these instances.

In contrast to the very high effective concentrations that are possible when interacting groups are held in the appropriate proximity and orientation, groups that are kept apart by the structure of the molecule of which they are a part have very low, or zero, effective concentrations. Intramolecular interactions are much more sensitive to their environment than are interactions between independent molecules in the liquid state.

Detailed explanations of the values of the effective concentrations measured are usually complicated by the presence of some unfavorable steric or physical interactions in the molecules with or without the interaction. Any strain in a molecule that is relieved upon forming the intramolecular interaction will increase the effective concentration. Consequently, there is no ideal example with which to illustrate the solely entropic contribution to the effective concentration, but the many experimental examples available indicate that the effect is very substantial.

Effective molarities for intramolecular reactions. A. J. Kirby (1980) *Adv. Phys. Org. Chem.* **17**, 183–278.

Dependence of effective molarity on linker length for an intramolecular protein–ligand system. V. M. Krishnamurthy *et al.* (2007) *J. Am. Chem. Soc.* **129**, 1312–1320.

2.5.B. Multiple Interactions: Entropy and Cooperativity

Knowledge of the covalent structure is usually sufficient to infer the chemical properties of small molecules, but not for macromolecules; their large sizes enable them to fold back on themselves so that many interactions can take place simultaneously among different parts of the molecule. A complex three-dimensional structure can result, which provides the unique environments and orientations of functional groups that give many macromolecules their special biological properties. Multiple groups on a single molecule can behave very differently than when in isolation.

For example, electrostatic, hydrogen bond and van der Waals interactions between two molecules are not particularly favorable energetically within an aqueous environment (Table 2-5) because there are competing interactions with the water surrounding the molecules. **For two molecules to interact favorably, they must overcome a loss of entropy, to fix one molecule relative to the second, and they must also interact with each other more strongly than they do individually with their environment**. For both reasons, interactions between individual molecules in solution are generally relatively weak (Table 2-5). For comparison, small molecules that interact with each other equally as well as they do with water would be expected to have $K_{AB} = 1/55\ \text{M} = 0.02\ \text{M}^{-1}$, where 55 M is the concentration of water molecules in liquid water. Only some of the values given in Table 2-5 are greater than this, and they are of molecules that are oppositely-charged or nonpolar, which would be expected to interact primarily via van der Waals interactions.

On the other hand, a polyelectrolyte with a number of such charged groups on a single molecule will bind ions of the opposite charge very tightly, owing to interactions between the charged groups. Being part of the same molecule, the charged groups are constrained by the covalent bonds to be close to each other, which is unfavorable energetically when they are of the same sign. To minimize this electrostatic repulsion, they attract counterions from the solution very tightly, depending upon the charge density of the polyelectrolyte and the valence of the counterions.

Table 2-5. Association in water of small molecules typical of noncovalent interactions in nucleic acids and proteins

Type of interaction	Example	Association constant (M^{-1})
Salt bridge	Acetate • guanidinium	0.37
	Acetate • amine	0.31
	Phenoxide • amine	0.20
Hydrogen bond[a]	Formic acid dimers	0.04
	Urea dimers	0.04
	N-methylacetamide dimers	0.005
	δ-Valerolactam dimers	0.013
van der Waals	Benzene dimers	0.4
	Cyclohexane • Cyclohexanol	0.9
	Benzene • Phenol	0.6

[a.] Interactions other than hydrogen bonding may contribute to the dimerization of these molecules, so the association constants are maximum values for hydrogen bonding.

From T. E. Creighton (1993) *Proteins: structures and molecular properties,* 2nd edn, W. H. Freeman, NY, p. 155.

Besides attracting counterions, another means of compensating for unfavorable electrostatic repulsions in a polyelectrolyte is to suppress the ionization of a fraction of the groups. Consequently, groups on a polyelectrolyte may have pK_a values very different from those found when they are isolated. These electrostatic effects between multiple groups on a polyelectrolyte are especially important for nucleic acids, with their multiple phosphate groups.

Effective charges of polyelectrolytes in a salt-free solution based on counterion chemical potential. T.Y. Wang *et al.* (2005) *J. Phys. Chem.* **109**, 22560–22569.

A modified Poisson–Boltzmann model including charge regulation for the adsorption of ionizable polyelectrolytes to charged interfaces, applied to lysozyme adsorption on silica. P. M. Biesheuvel *et al.* (2005) *J. Phys. Chem.* **109**, 4172–4180.

Similarity of salt influences on the pH of buffers, polyelectrolytes, and proteins. A. E. Voinescu *et al.* (2006) *J. Phys. Chem.* **110**, 8870–8876.

2.5.C. Cooperativity of Multiple Interactions: the Key to Macromolecule Folding

The simultaneous presence of multiple interactions within a single molecule will produce cooperativity between them, and together they can be much stronger than might be expected from the sum of their individual strengths. This is essential for macromolecules to adopt specific folded conformations, where the noncovalent interactions that can occur are individually very weak (Table 2-5). **Only by cooperating can a number of such simultaneous interactions produce a stable single conformation.** The following scenario is inspired by experimental measurements of disulfide bond stability during protein folding coupled to disulfide formation.

Consider an unfolded polymer chain in which there are two groups, A and B, capable of interacting favorably, as in a hydrogen bond, a salt bridge or a nonpolar hydrophobic interaction:

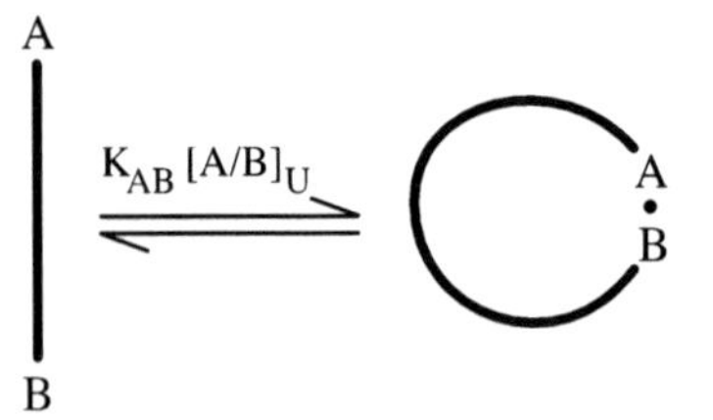

(2.17)

The observed equilibrium constant for interaction of the two groups, $K_{obs,U}$, is given by:

$$K_{obs,U} = K_{AB}[A/B]_U \quad (2.18)$$

where K_{AB} is the association constant measured with groups A and B on individual molecules (Table 2-5) and $[A/B]_U$ is the effective concentration of the two groups relative to each other on the unfolded polypeptide chain, U. Groups attached to moderate-sized random polymers have effective concentrations that have been measured experimentally to be in the region of 10^{-2}–10^{-5} M, depending upon their relative positions in the chain. With typical measured values of 0.01–1 M^{-1} for K_{AB} (Table 2-5), values for the observed equilibrium constant, $K_{obs,U}$, of between 4×10^{-3} and 10^{-7} are expected for individual hydrogen bonds, salt bridges, etc. Consequently, **a single interaction between two groups on a random polymer chain will be unstable** and present in only a small fraction of the molecules, unless the groups are close in the covalent structure in some way in which they have an especially high effective concentration.

Multiple interactions between two or more pairs of groups attached to the same molecule are usually not independent, because they can assist or interfere with each other. With two pairs of groups on a polypeptide, the following equilibria are possible:

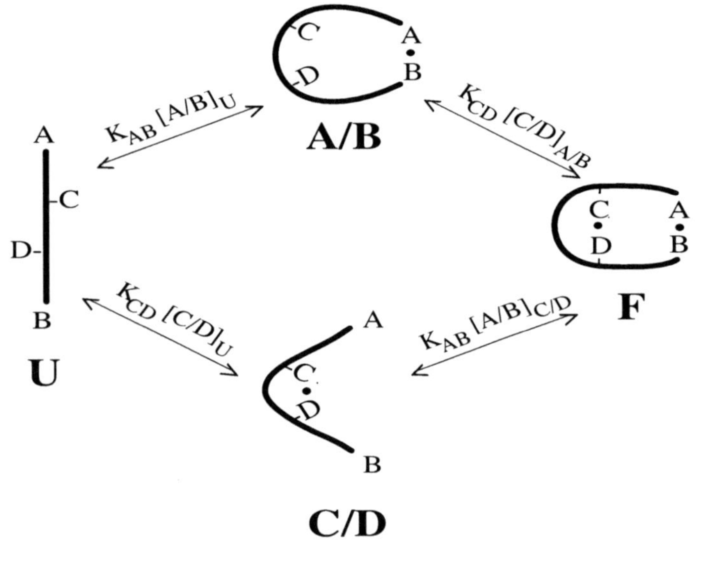

(2.19)

If both interactions are possible simultaneously, the presence of one interaction will increase the effective concentration of the other two groups. This occurs in a mutual manner, with both interactions having the same effect on each other, because the free energy change around any cycle must be zero and the two must be **linked functions** (Section 1.2.B):

$$\frac{[A/B]_{C/D}}{[A/B]_U} = \frac{[C/D]_{A/B}}{[C/D]_U} = \text{Coop} \tag{2.20}$$

The factor 'Coop' gives the degree of interaction between the two interactions. Consequently, a second simultaneous interaction can be more stable than when it is present alone, simply because it needs to lose less conformational entropy than in the absence of the first interaction, which tends to bring the second pair closer into proximity for interacting.

If additional groups that may also interact simultaneously are also present on the polymer chain, the above equilibria are extended in a similar way. The overall equilibrium constant between the final state, with all the interactions present, and the unfolded state, with none, is the product of the individual equilibrium constants along any of the conceivable reaction paths. For example:

$$K_{net} = (K_{AB}[A/B]_U)(K_{CD}[C/D]_{A/B})(K_{EF}[E/F]_F) \ldots \tag{2.21}$$

The value of K_{net} is independent of the reaction path, so we need not know or propose a specific 'folding pathway'.

The final state will be stable, i.e. populated by most of the molecules, only if the value of K_{net} is greater than unity. With weak interactions, the first will be very weak, with an equilibrium constant of between 10^{-3} and 10^{-7}. The presence of the first interaction can increase the effective concentration of the second pair of groups, and the equilibrium constant for the second interaction may be somewhat larger than that of the first, by the factor Coop. If the second equilibrium constant is still less than unity, however, the product of the two equilibrium constants will be even smaller than the first (Figure 2-8). Similarly, the net stabilities of conformations with additional weak interactions will be even lower than that of the conformation with a single interaction. This will continue until the effective concentrations of additional interacting groups are increased sufficiently to make each equilibrium constant for an additional interaction greater than unity. The value of K_{net} will then increase in magnitude with each additional interaction. With a sufficient number of weak interactions present simultaneously, the values of K_{net} will become greater than unity, and the folded conformation will be stable.

A highly idealized example with Coop = 10 for each additional interaction is described in Figure 2-8. Partially folded structures, with incomplete stabilizing interactions, will be unstable relative to the initial and final states, which means that the transition is cooperative. Each intermediate can be a mixture of a number of combinations of the 10 different interactions. A more realistic scheme would allow for the various intermediates to have varying free energies because of differences in their conformational properties. Also, the degree of cooperativity will be even greater if the intermediate structures have nonbonded groups in unfavorable environments, such as polar groups in nonpolar environments without being hydrogen bonded.

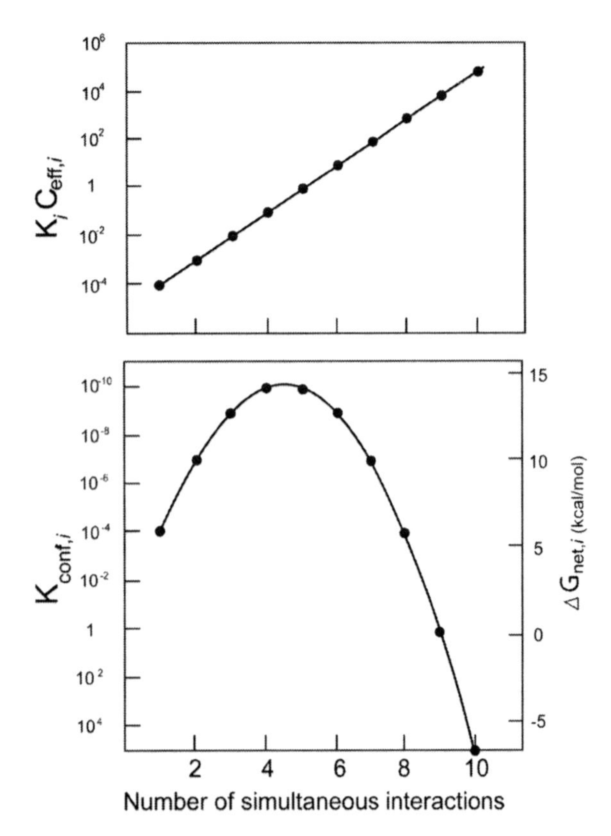

Figure 2-8. Hypothetical illustration of the cooperativity produced by multiple weak interactions. Up to 10 interactions are possible simultaneously, and the contribution of the *i*th interaction to the overall equilibrium constant is given at the *top*. The initial interaction has an equilibrium constant of 10^{-4}, and each additional interaction has an equilibrium constant that is 10 times greater than the previous one as a result of an increase in the effective concentration of the next groups to interact. The overall equilibrium constant $K_{conf,i}$ (*bottom*) is the product of the contributions of the *i* interactions present (Equation 2.21). Only with 10 such interactions is $K_{conf,i} > 1$, implying stability of the folded structure. The free energy of each state relative to U is given by $\Delta G^{\circ}_{conf,i} = -RT \log_e K_{conf,i}$, with the scale on the *right* pertaining to 25° C. Adapted from T. E. Creighton (1993) *Proteins: structures and molecular properties*, 2nd edn, W. H. Freeman, NY, p. 166.

This idealized scheme suggests that folded conformations could become infinitely stable by just increasing the number of interactions, but there is a physical limit to the number of interactions that can occur simultaneously on a single molecule. Also, the cooperativity arises because two groups interacting in a folded conformation do not need to lose the conformational freedom (entropy) that they had before interacting, and there is a limit to this entropic gain: it cannot be greater than the entropy that the interacting groups had in the first instance. Such considerations have been ignored here, simply to illustrate the principle of how cooperativity can occur.

Weak interactions are expected to stabilize a particular structure only when they cooperate so that the interacting groups have very high effective concentrations in that structure. The effective concentration of two groups in a folded structure depends upon the extent to which they are held in proximity when not interacting (Figure 2-9); in turn, this depends upon the stability of all the surrounding interactions. All parts of such a structure, therefore, are expected to be mutually dependent to varying degrees.

The contribution of each interaction to net stability of the folded structure should depend upon the effective concentration of the interacting groups in the folded structure (Figure 2-9). If the groups are on the surface, or in a flexible part of the structure, their effective concentration will be relatively low and the interaction will provide little, if any, net stability. Breaking that interaction will have little effect on the folded state. On the other hand, groups within relatively rigid parts of the structure will have high effective concentrations, and their interaction will provide a substantial contribution to the net stability. Removing or altering such an interaction would have a large effect on the stability of the folded conformation.

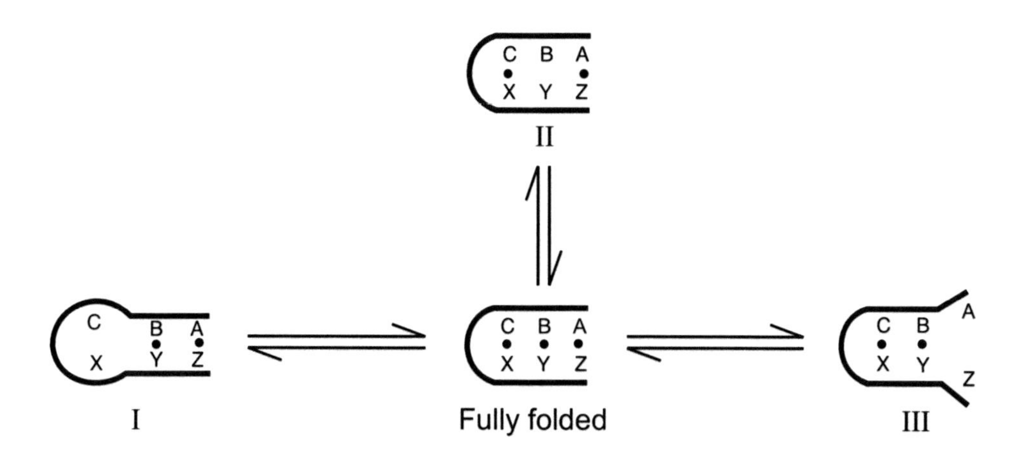

Figure 2-9. Simple schematic diagram of cooperativity among three simultaneous interactions occurring between groups A and Z, B and Y, and C and X in a macromolecule. The strength of each interaction is determined by the effective concentration of the two groups when they are not interacting, as in the variant conformations I, II and III. Assuming that there are no other considerations, the value of the effective concentration will be inversely proportional to the degree of flexibility permitted. Therefore, the most stable interactions should be those between groups that are held most rigidly by the other interactions, in this case B and Y, and that between A and Z would be expected to be the least stable. The stability of each interaction should depend upon the stabilities of all the others. Adapted from T. E. Creighton (1993) *Proteins: structures and molecular properties*, 2nd edn, W. H. Freeman, NY, p. 167.

Dissecting the roles of individual interactions in protein stability: lessons from a circularized protein. D. P. Goldenberg (1985) *J. Cell. Biochem.* **29**, 321–335.

Strategy for analysing the co-operativity of intramolecular interactions in peptides and proteins. A. Horovitz & A. R. Fersht (1990) *J. Mol. Biol.* **214**, 613–617.

Effective concentrations of amino acid side chains in an unfolded protein. K. Muthukrishnan & B. T. Nall (1991) *Biochemistry* **30**, 4706–4710.

~ CHAPTER 3 ~

AQUEOUS SOLUTIONS

The physical natures of the noncovalent interactions between atoms have been characterized, and are understood fairly well, with individual molecules in a vacuum or in a regular solid (Chapter 2), but the situation becomes much more complex in liquids. This is a consequence of the structural complexity of the liquid state, with its constantly changing interactions among many molecules in transient ensembles. Liquids are especially relevant to biology, where most macromolecules function only in an environment of liquid water or within membranes; the latter are formed as a result of the relatively poor interactions of lipids with water (Section 3.3). In spite of water's biological importance, and much study of it, it is not one of the best-understood liquids, and this limits our understanding of macromolecules immersed in it. The most important characteristic of all intermolecular forces between molecules dissolved in water is that these forces are often due more to the properties of this extraordinary solvent than to the intermolecular interactions themselves. The interactions of water with ions, dipoles and hydrogen bond acceptors or donors are so strong as to diminish greatly most of the forces that would occur between such groups in a vacuum or in a hydrocarbon solvent (Table 2-5). The strong interaction of the water molecules with each other produces a unique force between nonpolar atoms in water, the hydrophobic interaction (Section 3.2).

Pure water is not a physiologically relevant solution, and most solutions used in molecular biology contain many more components, especially salts and buffers; furthermore, other chemicals are often added in high concentrations to study the properties of proteins and nucleic acids, such as alcohols, urea, guanidinium chloride, etc. The presence of these additional **co-solvents** (or **co-solutes**) increases the complexity of even relatively simple solutions of a macromolecule, which must now be considered to consist of at least three components: (1) water, (2) macromolecule and (3) all co-solvents. Moreover, the amounts of these different components can change as a system changes because of chemical reactions or conformational changes that take place. When the co-solutes are present at high concentrations, the solutions are not ideal thermodynamically, because the co-solutes interact with each other, if only to take up physical space, and the classical **activity coefficients** normally used to describe nonideality can be far from unity.

This chapter will review the structure and properties of liquid water and will examine the various physical interactions in complex aqueous solutions with macromolecules.

The role of water in the thermodynamics of dilute aqueous solutions. R. A. Alberty (2003) *Biophys. Chem.* **100**, 183–192.

3.1. LIQUID WATER

The discussion of intermolecular interactions in Chapter 2 largely concentrated on interactions between pairs of molecules in a vacuum, where the nature of the interaction is relatively straightforward. In condensed media, in liquids, in solids and in macromolecules, numerous atoms and molecules are interacting simultaneously and usually inducing alterations in each other, so the exact treatment of all these interactions becomes much more problematic. For example, involvement of an O–H group as *donor* in a hydrogen bond increases the negative charge on the donor O atom, so it becomes a better hydrogen bond *acceptor* in a second hydrogen bond. van der Waals interactions between atoms affect their polarizabilities (Section 2.2.B.2) and the magnitude of the van der Waals interaction between two molecules is about 30% greater when they are part of a liquid than when in isolation in a vacuum.

Water? What's so special about it? J. L. Finney (2004) *Philos. Trans. Roy. Soc. Lond. B* **359**, 1145–1163.

Biological water: its vital role in macromolecular structure and function. F. Despa (2005) *Ann. N.Y. Acad. Sci.* **1066**, 1–11.

Do we underestimate the importance of water in cell biology? M. Chaplin (2006) *Nature Rev. Mol. Cell. Biol.* **7**, 861–866.

Water mediation in protein folding and molecular recognition. Y. Levy & J. N. Onuchic (2006) *Ann. Rev. Biophys. Biomol. Structure* **35**, 389–415.

3.1.A. Liquids: Close Interactions Without Order

Liquids are of central importance to both biology and chemistry, yet it is very difficult to describe them in detail. They have no readily defined structure, plus there is the complication of very many interactions occurring simultaneously between neighboring molecules. **Liquids are usually lighter than the corresponding solid by 5–15%, indicating increased distances, increased flexibility and decreased interactions between molecules.** Normal liquids have closely similar packing densities (Section 2.1.B) of between 0.48 and 0.61. These densities barely change with alterations in pressure, indicating that **the molecules are generally in van der Waals contact with their neighbors**; for example, doubling the atmospheric pressure generally decreases the volume of a liquid by only 0.01%. On the other hand, the liquid volume generally increases by about 0.1% for each °C rise in temperature, so **the molecules are also fluctuating substantially**.

The best experimental description of liquid structure comes from the scattering by X-rays or neutrons, which yields a **radial distribution function**, $g(r)$. This gives the symmetrically averaged density of atoms, relative to the bulk density, as a function of radial distance, r, from a reference atom (Figure 3-1). The value of $g(r)$ is zero at $r = 0$, and its value becomes substantial when r approaches twice the van der Waals radius of molecules that are roughly spherical, where neighboring atoms would be in close contact. At about this distance, $g(r)$ generally reaches a maximum, with a value indicating that **in a simple liquid 9 to 11 nearest-neighbor molecules are packed in approximately van der Waals contact around the central molecule and comprise the first shell of nearest neighbors**. The value of $g(r)$ then drops to a minimum at somewhat greater values of r, indicating that there are few spaces for molecules to penetrate the first shell of nearest neighbors. The density then increases to reach another maximum at a distance of just under two molecular diameters, corresponding to the second shell of neighboring atoms. The second maximum is markedly lower than the first, indicating that the order is diminishing with increasing distance. This trend continues, and a third shell may be apparent, but the atom density rapidly approaches that of the homogeneous bulk liquid. **The lower the temperature, the greater the degree of order in liquids**, unless the volume is kept constant by altering the pressure. More detailed descriptions of liquids come from numerical simulations, but their validity depends upon that of the model used for the calculations.

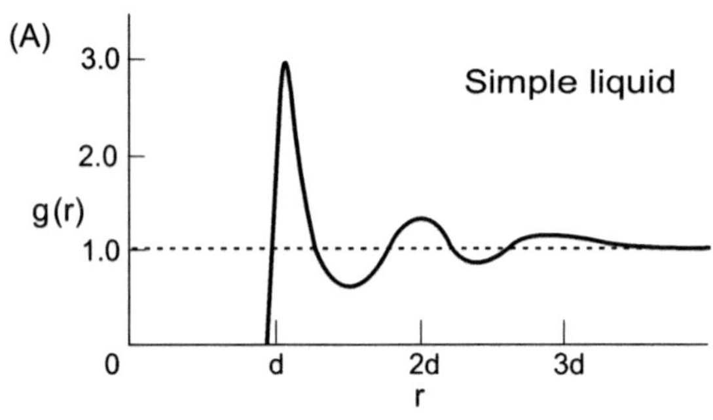

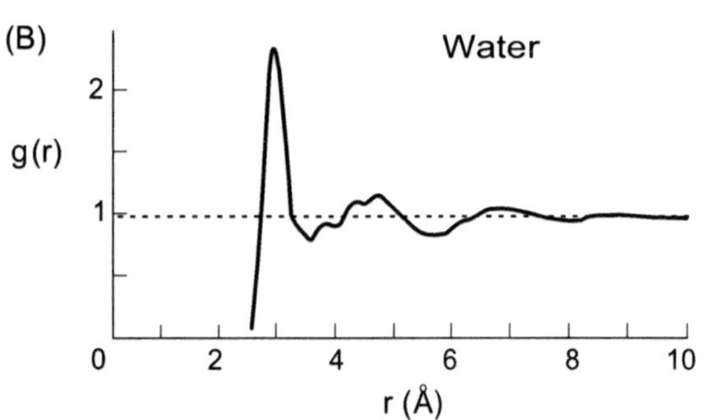

Figure 3-1. Radial distribution function of (A) a normal liquid and (B) water at 4° C. The probability of finding another atom at radial distance r from a reference atom is given by $g(r)$ times the density of the liquid. The distance r is expressed in (A) in terms of the van der Waals diameter of the molecule, d. The experimental curve for water measured using X-rays is given in (B); X-rays are scattered primarily by the O atoms of water molecules. Data from A. H. Narten & H. A. Levy.

Current models of liquids depict them as having close packing of hard-sphere molecules that is both irregular and constantly changing. **The shapes of the molecules and the harsh repulsive forces between them largely determine the properties of a liquid.** The structure of liquid argon is well represented by a box of marbles; liquid benzene is comparable to a box of Cheerios™ breakfast cereal. Although the attractive interactions between molecules stabilize the liquid phase, they play a minor role in determining its structure, unless they include hydrogen bonds or strong ionic interactions; this is the case with the most important liquid, water.

Generating inherent structures of liquids: comparison of local minimization algorithms. C. Chakravarty *et al.* (2005) *J. Chem. Phys.* **123**, 206101.

Improved evaluation of liquid densities using van der Waals molecular models. D. Mathieu & J. P. Becker (2006) *J. Phys. Chem.* **110**, 17182–17187.

Thermodynamic properties of van der Waals fluids from Monte Carlo simulations and perturbative Monte Carlo theory. A. Diez *et al.* (2006) *J. Chem. Phys.* **125**, 074509.

On the radial distribution function of a hard-sphere fluid. M. Lopez de Haro *et al.* (2006) *J. Chem. Phys.* **124**, 236102.

New evaluation of reconstructed spatial distribution function from radial distribution functions. D. Yokogawa *et al.* (2006) *J. Chem. Phys.* **125**, 114102.

Radial distribution function of penetrable sphere fluids to the second order in density. A. Santos & A. Malijevsky (2007) *Phys. Rev. E* **75**, 021201.

3.1.B. Water: the Importance of Hydrogen Bonding

The H_2O molecule is unique in having the same number of H atoms and of lone pair electron acceptors for hydrogen bonds. As a result, the numbers of hydrogen bond donors and acceptors are equal, and hydrogen bonding between water molecules is preeminent. The H_2O molecule has a bent geometry, with an O–H bond length of 0.957 Å and a bond angle of 104.5°:

O 0.957 Å

H 104.5° H (3.1)

With van der Waals radii of 1.2 Å and 1.45 Å for the H and O atoms, respectively, the molecule has a volume of 17.7 Å^3. The molecule in the stick representation given above appears very asymmetric, but the O atom has eight electrons and a share of the single electron of each hydrogen, so the electronic structure of the molecule is not very far from spherical (Figure 3-2); **the water molecule is often represented as a sphere of radius 1.4 Å.** The net charge is distributed asymmetrically, however, with excess electrons on the more electronegative O atom (Figure 3-2). Consequently, **the water molecule has a dipole moment of 1.85 Debye units, and the O–H bond can be considered to have 33% ionic character.** Note that the positive charge on the H atoms is located near the surface of the molecule, whereas the negative charge is closer to the center, so the positive partial charge can get closer to other atoms and interact more strongly. **Water appears to be a better donor for hydrogen bonds than acceptor.**

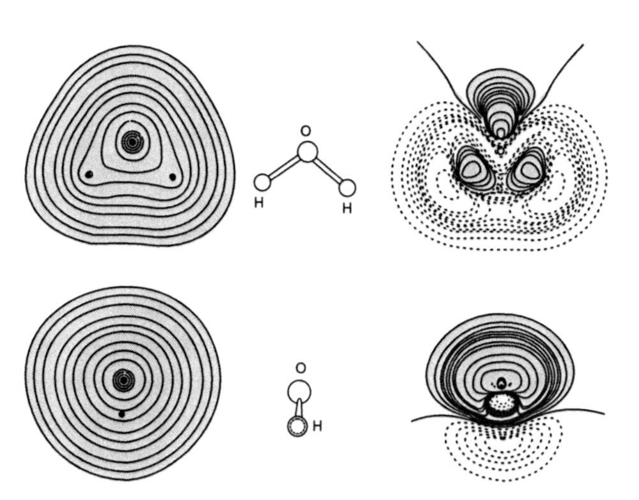

Figure 3-2. The electronic structure of the water molecule, shown as contour maps of electron density through the center of the molecule, viewed from two perpendicular angles. The total electron density is shown on the *left*, illustrating the nearly spherical shape of the molecule. The difference between this total electron density of the molecule and the density that would result from the superposition of individual spherical atoms is shown on the *right*. This illustrates the effect of covalent bonding on the electron density. The shift of electrons to the O atom is indicated by the positive electron density (*solid curves*) on the O atom and the negative electron density (*dashed lines*) on the H atoms. Data from I. Olovsson.

The predominance of hydrogen bonding for determining the properties of water is amply illustrated by the structure of ice (Figure 3-3). The structure gives the impression of being determined exclusively by the four hydrogen bonds that each H_2O molecule participates in, two as hydrogen donor, two as acceptor. The angle between the H atoms in the water molecule (104.5°) is very close to that ideal for the tetrahedral packing, 109.5°. **Due to the hydrogen bonding, the crystal structure of ice is much more open than that of a more typical crystal.** Each molecule has only four immediate neighbors, instead of the usual 12 in crystalline close-packing of spheres. Only 42% of the volume is filled by the van der Waals volume of the molecules, rather than the 74% observed with spherical close-packing.*

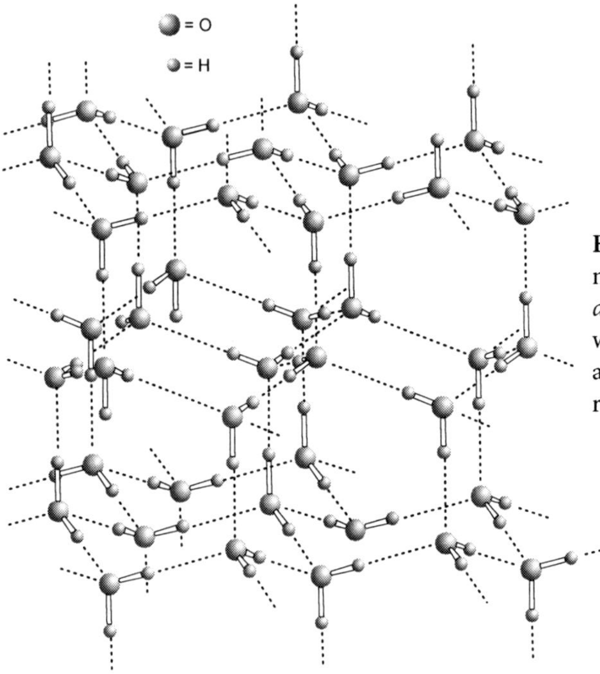

Figure 3-3. Structure of normal ice. Each H_2O molecule is involved in four hydrogen bonds (*thin, dashed lines*), each 2.76 Å between O atoms. The water molecule is donor in two hydrogen bonds, acceptor in the other two. Substantial empty channels run between the molecules.

The four hydrogen-bonded water molecules have their O atoms 2.76 Å from the central O atom, with the next-nearest neighbors 4.5 Å away. The hydrogen bonds in ice are 0.2 Å shorter than those between isolated H_2O dimers (2.98 Å), presumably because of cooperativity: participation of the H atom in a hydrogen bond as a donor causes the O atom to be much more effective as an acceptor in a second hydrogen bond, and vice versa. There may also be a contribution from the order of the crystal, with so many hydrogen bonds being present simultaneously (Section 2.5.C).

The partial negative charge on the O atom is frequently described as being localized primarily on the two lone-pair electrons that effectively project above and below the plane of the molecule, giving water a tetrahedral structure (Figure 3-2). Hydrogen bonding in ice and in H_2O dimers in the gas phase directs the H atoms towards the positions of the lone-pair electrons, but this geometry is now believed to result primarily from repulsions between the H atoms. The lone-pair electrons on the O atom are smeared out between the tetrahedral positions, lying in the plane bisecting the molecule (Figure 3-2). There is probably only a slight energetic preference for a tetrahedral arrangement of hydrogen bonds, and that apparent in ice probably results primarily from packing considerations.

* *The hydrogen bonds of ice decrease the van der Waals volume of each water molecule by 4.0 Å^3. If this were not taken into account, the density of ice would have suggested a packing volume of 54%.*

The structure of liquid water is still uncertain, in spite of many experimental and theoretical studies, and no single model can explain all of its properties. The unusual properties of water (Table 3-1) are well known, particularly its anomalous density change with temperature, when it expands at decreasing temperatures below about 4° C. The radial distribution function of water also shows substantial differences from those of other liquids (Figure 3-1). That measured with X-rays reveals primarily the relative positions of the O atoms. The value of $g(r)$ is zero for r less than 2.5 Å, but rises to a maximum at 2.82 Å at low temperatures. The position of this maximum is just slightly greater than the distance between hydrogen-bonded neighbors in normal ice, and this distance increases only somewhat at higher temperatures, up to 2.94 Å at 200° C. The number of nearest neighbors is approximately 4.4, greater than the 4.0 in ice, but much lower than the mean value of about 10 in most liquids. A second maximum occurs at 3.5 Å, a distance that does not coincide with any in ice, but this peak is now thought to be an artifact of the diffraction analysis. The next maximum is at about 4.5 Å, which corresponds to the distance between pairs of O atoms that are hydrogen bonded in ice to the same water molecule. A further maximum occurs at about 7 Å, which would be the next-nearest neighbor in ice. After that, there is little evidence for further order. The maxima at 4.5 and 7 Å largely disappear at temperatures greater than 50° C, indicating a breakdown of even the local structure, and the number of nearest neighbors increases to about 5 at 100° C.

Compared with other liquids of molecules of a similar small size, water has higher melting and boiling temperatures, and greater values for the dielectric constant, heat of fusion, heat of vaporization, specific heat capacity and surface tension (Table 3-1). These properties all indicate that the **molecules interact more strongly with each other in water than do those of other liquids**, presumably due to the hydrogen bonds, and indicate that liquid water retains much of the cohesive energy of ice. The X-ray and neutron-scattering data confirm the importance of hydrogen bonding for the structure of water, but how a disordered, liquid state results is still not entirely clear. Part of the structural disorder may arise from flexibility in the water molecule itself, both in the variation of O–H bond lengths and in an average variation of about ±15° in the normal H–O–H bond angle of 104.5°; as a result of this intrinsic flexibility, water can form no fewer than 12 different ice structures. The local tetrahedral arrangement of four hydrogen-bonded near-neighbors of ice appears to persist in liquid water, but with a fifth neighbor also frequently present. In contrast to ice, where the rotation about the hydrogen bond is limited to one of three angles by the crystal lattice, very many orientations of neighboring molecules probably exist in the liquid, so the relative positions of H atoms on neighboring molecules are not well defined.

The dielectric constant of water (and of liquids made up of other molecules containing dipoles) is so large because the electrons of the water molecules can redistribute in the presence of a charged atom. In addition, the water molecules themselves can redistribute. They tend to line up so that their dipoles are oriented in line with the electric field.

High pressures cause compression and increased packing of normal liquids, which slow down all molecular motions. In water, on the other hand, the mobility increases with initial compression, and only further increases in pressure restrict the motional freedom. This might indicate that the initial increase in pressure causes a breakdown or distortion of the tetrahedral hydrogen bonding between molecules, liberating them for greater mobility.

Table 3-1. Physical properties of water at 25° C

Property	Value
Molecular weight	18.01529 Da
O–H bond length	0.9572 Å
H–O–H bond angle	104.52°
Effective diameter	2.75 Å
Normal vibrations as gas	
Symmetric stretching	3656.65 cm^{-1}
Deformation	1594.59 cm^{-1}
Asymmetric stretching	3755.79 cm^{-1}
Dipole moment, gas	1.834 D
Dipole moment, liquid	2.45 D
Polarizability	1.470×10^{-30} m^3
Dielectric constant	78.54
Density	0.997045 g/cm^3
Packing density	36.29%
Coordination number	4.4
Heat of fusion (0° C)	6.008 kJ/mol
Heat of vaporization	43.991 kJ/mol
Specific heat capacity	4.1796 J K^{-1} g^{-1}
Vapor pressure	3.1675 kPa
Surface tension	71.96 mJ/m^2
Refractive index	1.33287
Thermal expansivity	2.5721×10^{-4} K^{-1}
Isothermal compressibility	0.452472 GPa^{-1}
Translational diffusion coefficient	2.14×10^{-9} m^2/s
Rotational diffusion coefficient	6.06×10^{10} s^{-1}
Viscosity	0.8904 mPa s
Ionic product	1.0×10^{-14} M

Neutron scattering analysis reveals the relative positions of the H atoms. Those in liquid water are not fixed beyond about 5 Å, suggesting that the water molecules rotate freely about the hydrogen bonds, at a picosecond time scale, in contrast to ice, where the tetrahedral crystal lattice keeps them fixed.

Many models of liquid water have all the molecules hydrogen-bonded all of the time, but with a great variety of hydrogen-bond geometries and energies. Others have each group hydrogen-bonded only a fraction of the time. Some models incorporate the experimental suggestions that water is a mixture of two states in equilibrium. One state is envisaged to have a relatively low enthalpy, low entropy (Section 1.5) and large volume, similar to hydrogen-bonded ice. The other has relatively high enthalpy, high entropy and small volume, analogous to a normal liquid with much less hydrogen bonding. At the present time, no particular model seems obviously more realistic than all the others.

In spite of uncertainty about the precise nature of liquid water, the strong hydrogen bonding between water molecules is clearly the basic explanation of many of the peculiar properties of this solvent. It also causes thermodynamic studies of phenomena in water to be particularly complex, because changes in entropy and enthalpy tend to be mutually compensating, with relatively little change in free energy, a phenomenon known as **enthalpy–entropy compensation** (Section 1.5). For example, formation of hydrogen bonds in water should produce a favorable decrease in enthalpy, ΔH, but requires an unfavorable decrease in entropy, ΔS, because the molecules participating in the hydrogen bonds must be relatively fixed in orientation and proximity. These two contributions to the Gibbs free energy, G, tend to cancel out, because $\Delta G = \Delta H - T\Delta S$, where T is the temperature. Even with large changes in enthalpy and entropy, relatively little or no change in free energy may result. Other, relatively small, effects may thus predominate in determining the free energy of any such transition in water. Rationalization of thermodynamic data in water is very tricky indeed.

Energetics of hydrogen bond network rearrangements in liquid water. J. D. Smith *et al.* (2004) *Science* **306**, 851–853.

Ultrafast memory loss and energy redistribution in the hydrogen bond network of liquid H_2O. M. L. Cowan *et al.* (2005) *Nature* **434**, 199–202.

Hydrogen bonds in liquid water are broken only fleetingly. J. D. Eaves *et al.* (2005) *Proc. Natl. Acad. Sci. USA* **102**, 13019–13022.

Ordering of hydrogen bonds in high-pressure low-temperature H_2O. Y. Q. Cai *et al.* (2005) *Phys. Rev. Lett.* **94**, 025502.

Unified description of temperature-dependent hydrogen-bond rearrangements in liquid water. J. D. Smith *et al.* (2005) *Proc. Natl. Acad. Sci. USA* **102**, 14171–14174.

Correlation of hydrogen bond lengths and angles in liquid water based on Compton scattering. M. Hakala *et al.* (2006) *J. Chem. Phys.* **125**, 084504.

Multidimensional infrared spectroscopy of water. II. Hydrogen bond switching dynamics. J. J. Loparo *et al.* (2006) *J. Chem. Phys.* **125**, 194522.

Relevance of hydrogen bond definitions in liquid water. M. Matsumoto (2007) *J. Chem. Phys.* **126**, 054503.

3.2. THE HYDROPHOBIC INTERACTION: AVOIDING WATER'S PHOBIA

Water is a very poor solvent for nonpolar molecules, compared with normal organic liquids. Nonpolar molecules cannot participate in the hydrogen bonding that appears to be so important in liquid water, and aqueous solutions of such molecules have many anomalous physical properties. This relative *absence* of a favorable interaction between nonpolar molecules and water causes the alternative favorable interaction between the nonpolar groups themselves to be much more important than would be the case in a normal solvent. This preference of nonpolar atoms for nonaqueous environments has come to be known as the **hydrophobic interaction**. It is a major factor in the stabilities of proteins, nucleic acids and membranes (Section 3.3) and it also has some unusual characteristics.

Stability of protein structure and hydrophobic interaction. P. L. Privalov & S. J. Gill (1988) *Adv. Protein Chem.* **39**, 191–234.

Structure and bonding of molecules at aqueous surfaces. G. L. Richmond (2001) *Ann. Rev. Phys. Chem.* **52**, 357–389.

Molecular theory of hydrophobic effects: 'she is too mean to have her name repeated'. L. R. Pratt (2002) *Ann. Rev. Phys. Chem.* **53**, 409–436.

Hydrophobic effects and modeling of biophysical aqueous solution interfaces. L. R. Pratt & A. Pohorille (2002) *Chem. Rev.* **102**, 2671–2692.

Hydrophobic interaction and hydrogen-bond network for a methane pair in liquid water. J. L. Li *et al.* (2007) *Proc. Natl. Acad. Sci. USA* **104**, 2626–2630.

3.2.A. Partition Coefficients: Measuring Preferences for Different Environments

The tendency of a molecule to prefer one environment over another is measured by its partition coefficient between the two environments, which can be a gas and a liquid or two immiscible liquids, such as water and cyclohexane. The two environments are brought into contact and equilibrated, and the relative concentrations of the molecule in the two are measured. This gives an equilibrium constant for the partitioning, which is known as the **partition coefficient**. For example, the relative concentrations, [X], of the molecule X in the vapor phase and in the aqueous phase at equilibrium (after sufficient stirring or shaking) give the partition coefficient, K_D, for the transfer of X from vapor to water:

$$K_D = \frac{[X]_{water}}{[X]_{vapor}} \tag{3.2}$$

The **free energy of transfer** is derived from the partition coefficient by:

$$\Delta G_{tr} = -RT \log_e K_D \tag{3.3}$$

This free energy is a measure of the relative free energies of the molecule in the two environments. A molecule in the vapor state has no interactions with other molecules, so the free energy of transfer is primarily a measure of the interactions of the molecule with water. A negative free energy of transfer would imply in this case that the molecule is more stable in water than in isolation in the vapor state. Its preference for the aqueous phase is a measure of the **hydrophilicity** of the molecule. In contrast, the tendency of a molecule to enter a nonpolar solvent from water is a measure of its **hydrophobicity**.

To check for complications such as aggregation of the molecule in one of the phases, the measurements must be made at a series of concentrations and the results extrapolated to infinite dilution. For example, very polar molecules might associate in a nonpolar solvent to pair up their polar groups, whereas very nonpolar molecules might aggregate in water to minimize their exposure to it. Aggregation in one phase will cause the equilibrium constant to shift in its favor with increasing concentrations of the molecule. Also, very polar groups may 'drag' hydrogen-bonded water molecules into the nonpolar phase, and this should be checked using tritiated water (Section 5.5.A.1).

The solubility of a molecule in a liquid is determined by the energetic cost of making a cavity of the appropriate size in the liquid, compensated by the extent to which favorable interactions occur between the molecule and those of the liquid. In the vapor phase, in contrast, there is no cavity and no interactions with the environment. Consequently, a partition coefficient of unity for transfer from a liquid to the vapor phase implies that the energetic cost of making a cavity in the liquid for the molecule is exactly compensated by its favorable interactions with the liquid.

The hydrophilicities of molecules are quite straightforward. **The more polar groups on the molecule, the greater the hydrophilicity** (Table 3-2). The least hydrophilic molecules are saturated hydrocarbons. The most hydrophilic molecules are those with charged groups. Ionization of the molecules in Table 3-2 increases their hydrophilicities dramatically. The interactions of charged groups with water are so strong that it is difficult to remove them from it. This was a major factor in limiting the usefulness of mass spectrometry with biological molecules, until the advent of special methods to introduce them into the vapor phase (Chapter 6). Ionized molecules also do not cross membranes readily.

The partition coefficients of moderately complex molecules can often be estimated or predicted from the partition coefficients of the chemical groups that comprise the molecule. For example, adding an amino group to ethane (CH_3–CH_3) increases the hydrophilicity equilibrium constant by 5 orders of magnitude, so adding one to acetic acid, to produce the amino acid glycine, should have the same effect and increase its hydrophilicity equilibrium constant to about 10^{+10}.

Estimation of molecular linear free energy relationship descriptors by a group contribution approach. 2. Prediction of partition coefficients. J. A. Platts *et al.* (2000) *J. Chem. Inf. Comp. Sci.* **40**, 71–80.

Recent methodologies for the estimation of n-octanol/water partition coefficients and their use in the prediction of membrane transport properties of drugs. G. Klopman & H. Zhu (2005) *Mini Rev. Med. Chem.* **5**, 127–133.

A structural analogue approach to the prediction of the octanol-water partition coefficient. A. Y. Sedykh & G. Klopman (2006) *J. Chem. Inf. Model.* **46**, 1598–1603.

3.2.B. The Hydrophobic Interaction in Nonpolar Model Systems

The hydrophobic interaction results from the tendency of nonpolar molecules to interact with each other, rather than with water. The magnitude of the hydrophobic interaction is usually measured by the free energy of transfer (ΔG_{tr}, Equation 3.3) from water to the gas, liquid or solid states. The free energy of transfer of a nonpolar molecule from water is negative in each case, indicating **that the nonpolar molecule prefers a nonaqueous environment, even the vacuum**. This is strictly true only for aliphatic molecules; aromatic molecules are slightly more polar, probably because of their excess

Table 3-2. Hydrophilicities of simple small organic compounds in uncharged form

Class	Example	Log hydrophilicity
Guanidines	$CH_3-NH-C(=NH)-NH_2$	8.2
Amides	$CH_3-C(=O)-NH_2$	7
Diols	CH_2OH-CH_2OH	7
Peptides	$CH_3-C(=O)-NH(CH_3)$ $CH_3-C(=O)-N(CH_3)_2$	7
Phosphotriesters	$(CH_3O)_3P=O$	6.2
Carboxylic acids	$CH_3-C(=O)-OH$	5
Water	H_2O	5
Amines	$C_2H_5-NH_2$	3.5
Alcohols	C_2H_5-OH	3.6
Nitriles	CH_3-CN	2.8
Ketones, Aldehydes	$CH_3-C(=O)-(CH_3, H)$	2.7
Esters	$CH_3-C(=O)-O-CH_3$	2
Thioethers, ethers	CH_3-S-CH_3, CH_3-O-CH_3	1.3
Chlorides	C_2H_5-Cl	1.2
Thiols	C_2H_5-SH	1
Alkynes	$CH\equiv CH$	0
Alkenes	$CH_2=CH_2$	-1
Alkanes	CH_3-CH_3	-0.9

The hydrophilicity is measured by the equilibrium constant for partitioning of the molecule from vapor to an aqueous solution. Data from R. V. Wolfenden.

of π electrons within the center of the ring and the deficit around their edge (Section 2.2.B), and aromatic rings actually have slightly favorable interactions with water. Nevertheless, they interact more strongly with nonpolar molecules in the liquid or solid phases, so they are still only poorly soluble in water and participate in the hydrophobic interaction.

There is much debate as to what is the best nonpolar solvent to use to measure hydrophobicities. Octanol and chloroform are often used, but neither are totally nonpolar molecules, and both of their phases acquire substantial quantities of water after equilibration. Chloroform participates in hydrogen-bonding and contains about 0.1 M water at saturation. Cyclohexane appears to be the most suitable, as it has no polar groups and contains no more than 1 mM water after equilibration.

The **thermodynamics of transfer to water of nonpolar molecules are anomalous in being markedly temperature-dependent**, as illustrated in Figure 3-4 for a molecule the size of cyclohexane at two different temperatures. In considering the thermodynamics, keep in mind that the enthalpy change (ΔH) reflects the difference in the magnitude of the noncovalent interactions between molecules that occur in the two phases, while the entropy change (ΔS) reflects the difference in disorder of the system (Chapter 1). Transferring a solute molecule to a liquid involves (1) creating a suitable cavity in the liquid, (2) introducing the solute molecule into the cavity and then (3) rearranging the solute and the surrounding liquid molecules to optimize the interactions between them. The observed thermodynamic parameters of transfer are the net difference between the two phases of all three factors, so physical interpretation of thermodynamic parameters for transfer is not always straightforward. This is especially the case with water, where increased hydrogen bonding results in a more negative enthalpy, but also a more negative entropy due to the necessity of fixing the positions of the interacting molecules; the two have compensating effects on the free energy (Section 1.5). Nevertheless, analysis of the temperature-dependence of the hydrophobic interaction is crucial for understanding its physical basis.

The thermodynamics of transfer data indicate that it is **the aqueous solution containing the nonpolar molecule that has anomalous physical properties**. The differences in the thermodynamic parameters for transfer from the gas, liquid or solid states to water simply reflect the normal differences between these three physical states. For example, the nonpolar liquid has favorable van der Waals interactions between molecules that are essentially absent in the gas phase, but the liquid also has less disorder than the gas; these differences are apparent in the negative changes in both ΔH and ΔS upon transfer from the gas to the nonpolar liquid. Changes of the same type, but of smaller magnitude, occur upon solidification of the nonpolar liquid.

At room temperature, **the unfavorable transfer of a nonpolar molecule from a nonpolar liquid to water is observed to be primarily a result of the unfavorable change in entropy**. The enthalpy change is approximately zero at room temperature, so there are similar enthalpic interactions in the aqueous solution and in the nonpolar liquid. The precise temperature where $\Delta H_{tr} = 0$ is known as T_H (Figure 3-4). The unfavorable entropy change is believed to result from an increased ordering of the water molecules around the nonpolar molecule. These water molecules appear to be more tightly packed than those of normal bulk water, as the measured partial volumes of nonpolar molecules are smaller in water than in other liquids. Water molecules cannot make hydrogen bonds to a nonpolar solvent, so they are imagined to satisfy their hydrogen-bond potential by forming a hydrogen-bonded 'iceberg' network around the nonpolar surface. Extreme examples of such ordered water cages, known as **clathrates**, are observed around apolar gases dissolved in water at low temperatures and high gas pressures. The clathrate water molecules are fully hydrogen-bonded, as in ice, although with

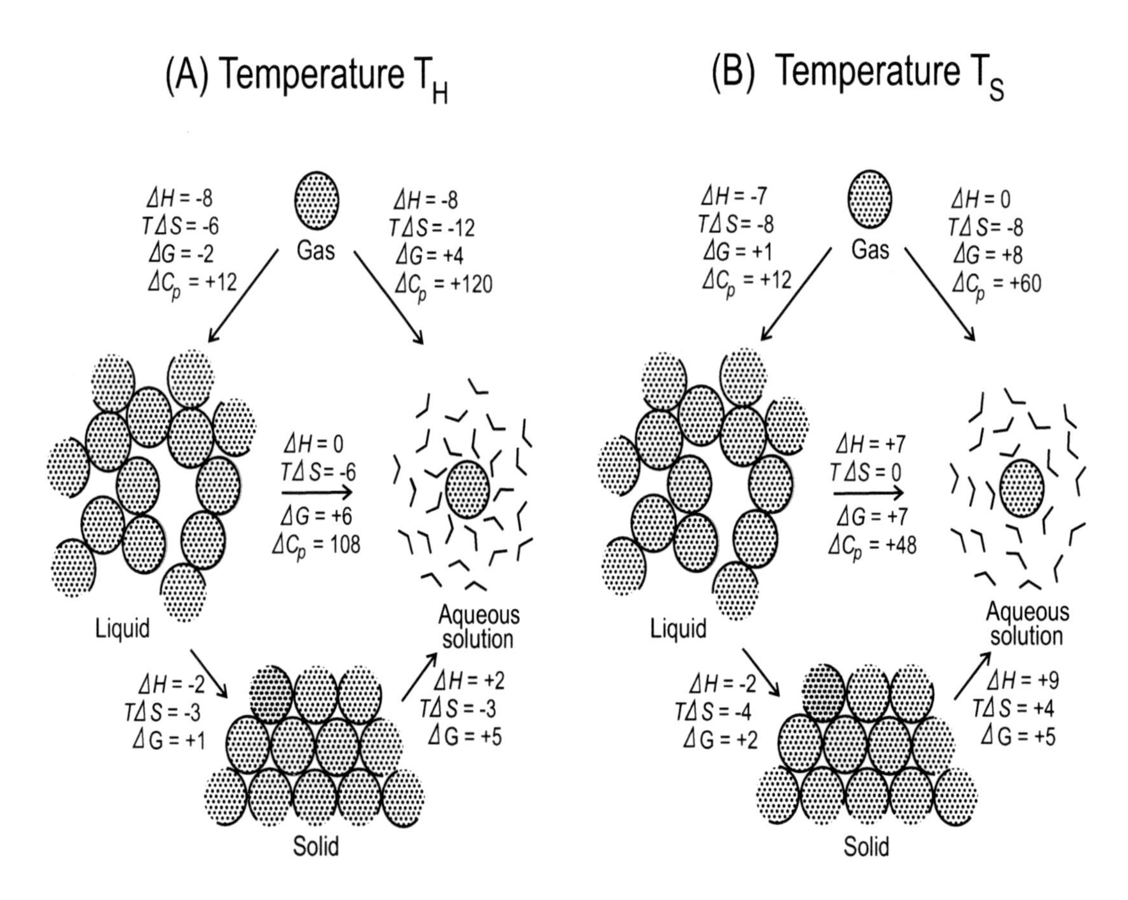

Figure 3-4. Typical thermodynamics of transfer of a nonpolar molecule the size of cyclohexane between the gas, liquid and solid phases and aqueous solution at temperatures (A) T_H, approximately 20° C, where $\Delta H_{tr} = 0$ for transfer between liquid and water, and (B) T_S, approximately 140° C, where $\Delta S_{tr} = 0$. The values of ΔH, TΔS and ΔG are in units of kcal/mol, that of ΔC_p in units of cal/°K mol. Adapted from T. E. Creighton (1993) *Proteins: structures and molecular properties*, 2nd edn, W. H. Freeman, NY, p. 157.

nonoptimal geometries. A similar ordering of water molecules around a nonpolar solute in aqueous solution is believed to occur under normal conditions, although to a lesser degree, as the water molecules are packed more densely. In doing so, the water molecules become more ordered and lose entropy, but their increased hydrogen bonding compensates by decreasing their enthalpy. Because the entropic factor dominates the unfavorable ΔG_{tr} to water, it was thought originally that the water-ordering effect is responsible for the low solubility of nonpolar molecules in water. Further analysis, however, demonstrates just the opposite: the **water-ordering tends to *increase* the solubility of the nonpolar molecule**.

As the temperature is increased, **the relatively ordered water shell around the nonpolar solute tends to melt out and to become more like bulk water**. This melting of the ordered water produces an anomalously large **heat capacity**, C_p, of such an aqueous solution (Section 1.5), and the large C_p is the thermodynamic hallmark of aqueous solutions of nonpolar molecules. It causes the thermodynamic parameters of such solutions to be markedly temperature-dependent (Figure 3-5), because the heat capacity defines the temperature-dependence of the enthalpy, entropy and free energy (Section 1.5). The value of **the heat capacity is generally found to be proportional to the nonpolar surface area of the solute molecule exposed to water, as are the other thermodynamic parameters** (Figure 3-6). The heat capacity is substantial, corresponding to about 20% of the normal heat capacity of the water molecules involved in the first hydration shell.

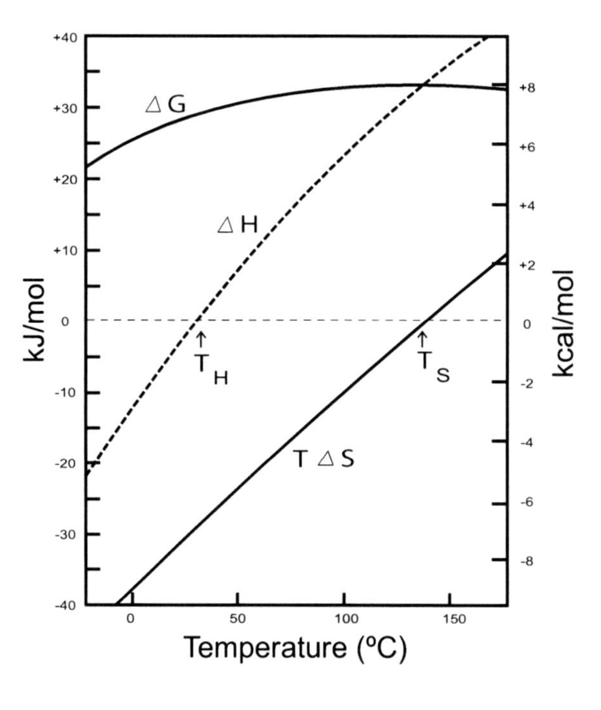

Figure 3-5. Typical thermodynamics of transfer of a hydrocarbon from the liquid to aqueous solution, using pentane as an example. The strong temperature-dependence of both the enthalpy and entropy difference between the two phases is a result of the different heat capacities of the two phases. The free energy difference is the net difference between the enthalpic and entropic contributions. It reaches a maximum where $\Delta S° = 0$, whereas the equilibrium constant (which is proportional to $-\Delta G°/T$) reaches a maximum where $\Delta H° = 0$. Data from P. L. Privalov & S. J. Gill.

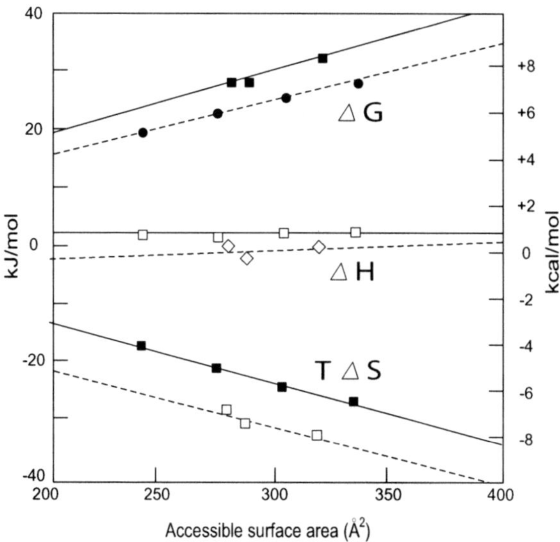

Figure 3-6. Thermodynamics of dissolution of hydrocarbon liquids into water at 25° C as a function of the accessible surface area of the hydrocarbon. The enthalpy change is virtually zero at this temperature. The *dashed lines* are for aliphatic molecules, the *solid lines* for aromatics. Kindly provided by S. J. Gill. Adapted from T. E. Creighton (1993) *Proteins: structures and molecular properties*, 2nd edn, W. H. Freeman, NY, p. 159.

The temperature-dependence of the hydrophobic interaction provides important clues to its physical nature. At temperatures above T_H, the entropy of transfer decreases and becomes less unfavorable for transfer to water, but the enthalpy change becomes more unfavorable. The entropy of transfer from the liquid to water becomes zero at temperature T_S (Figure 3-4). The value of T_S was originally considered to be about 110° C, when ΔC_p was thought to be independent of temperature. The value of ΔC_p is now known to decrease at higher temperatures, and T_S is believed to be about 140° C (Figure 3-5). However, the temperature-dependence of ΔC_p affects primarily extrapolations to very high, nonphysiological temperatures, and it is often convenient to approximate ΔC_p as constant.

The large changes in ΔH_{tr} and ΔS_{tr} with temperature largely compensate, and the value of ΔG_{tr} changes much less (Figure 3-5). Nevertheless, the magnitude of the hydrophobic interaction has a maximum value at one temperature and decreases at both higher and lower temperatures. Hydrophobicity measured as ΔG_{tr} is at a maximum at temperature T_S, whereas hydrophobicity measured by the equilibrium constant for transfer is at a maximum at temperature T_H. It is important, therefore, to define what measure of hydrophobicity is being used.

At the higher temperature, T_S, the value of zero for the ΔS_{tr} from liquid to water suggests that the net water-ordering effect has disappeared and that water is now a normal solvent. The difficulty with this interpretation is that the ΔC_p, which is believed to arise from the water-ordering, is still substantial, although diminished (Figure 3-4). Nevertheless, there appear to be no net interactions between a nonpolar solute molecule and water at this temperature, because the ΔH_{tr} from the gas to water is approximately zero (Figure 3-4). A possible explanation for this situation is that totally disordered water would have net repulsions with nonpolar solutes and that sufficient ordering persists at T_S to balance such repulsions. All the details are not clear, but the important point is that **water becomes a more normal solvent at high temperatures**.

At the high temperature T_S, in the apparent absence of the net water-ordering effect, the low solubility of nonpolar molecules in water is apparently due to the much poorer enthalpic interactions between the nonpolar molecules and water than between nonpolar molecules in the nonpolar liquid and between water molecules in liquid water. **The van der Waals interactions in a nonpolar liquid or solid are much greater than those between water and the nonpolar solvent. Likewise, the hydrogen bonds that exist in water cannot be formed with the nonpolar molecule and must be disrupted when it is dissolved in water.** These two phenomena are the basis of the hydrophobic interaction.

The very small size of the water molecule also seems to be important for the hydrophobic effect. For example, the equilibrium constant for entry of an isolated water molecule from the vapor phase into cyclohexane is almost exactly unity at room temperature, so the cost of making a cavity in a nonpolar solvent that is large enough to accommodate a single water molecule is just balanced by any net attraction between the water molecule and the walls of the cavity. Both factors may be very small, but in any case molecules smaller than water prefer the vapor phase, whereas molecules larger than water prefer cyclohexane.

The hydrophobic interaction results from a preference of nonpolar atoms to interact with each other, rather than with water. This interaction has the unusual property of its equilibrium constant increasing in magnitude at high temperatures. This is a result of the tendency of water to form ordered hydrogen-bonded networks around the nonpolar molecule at low temperatures, which decreases at higher temperatures. The water-ordering effect increases the solubility of nonpolar molecules in water, and it seems to be water's attempt to improve its interactions with the nonpolar molecule. The water-ordering effect is not responsible for the poor solubility of nonpolar molecules, as is often assumed from the way the entropy change dominates the thermodynamics of transfer at room temperature. Instead, **the water-ordering effect is responsible for the decrease in the magnitude of the hydrophobic interaction at low temperatures**. It is important, therefore, not to use terms like 'hydrophobicity', 'hydrophobic interaction' or 'hydrophobic effect' to refer to the water-ordering effect or to the resultant anomalous thermodynamic parameters. The water-ordering effect increases the solubility of nonpolar molecules in water and has opposite implications to the usual meaning of the term 'hydrophobic interaction'. Although the exact nature of the water-ordering that occurs in the solvation of nonpolar surfaces by water is still uncertain, it is the primary cause of the complex thermodynamics of the hydrophobic interaction.

To summarize, the hydrophobic interaction arises because nonpolar molecules have much stronger interactions with each other than with water. It is due to a combination of the relative absence of favorable interactions between water and nonpolar molecules and of the much more favorable van der Waals interactions between nonpolar molecules (Figure 3-4). The thermodynamics of the hydrophobic interaction are complex because water attempts to minimize its unfavorable interactions with nonpolar surfaces by forming more ordered water structures around them. These water structures are melted out at high temperatures, however, and are more prevalent at low temperatures. The more favorable interactions between nonpolar molecules and water at low temperatures decrease the magnitude of the hydrophobic interaction and cause the values of the enthalpy and entropy of the interaction to be temperature-dependent, with a large heat capacity. Note that $^{2}H_2O$ (D_2O) is a poorer solvent for nonpolar amino acids than H_2O, implying that the hydrophobic effect is greater in D_2O.

The hydrophobic interaction as measured in model systems, by the partition coefficient of a nonpolar molecule between a nonpolar solvent and water, is a balance between two factors: van der Waals interactions in the nonpolar liquid and hydration in water. When considering the role of the hydrophobic interaction in biological systems, it may be better to separate the two.

Large molecules in water can also interact to varying extents over a significant distance, without coming into contact and forming a complex, by altering the water structure between them. For example, large flat nonpolar surfaces have been shown to interact in water over distances as great as 25 Å. The interaction energy does not vary gradually with the distance, but exhibits oscillations with an average periodicity of 2.5 Å, approximately the diameter of a water molecule. The atomic structure of the water between the surfaces largely determines the interaction between them; **the most favorable interactions occur when the distance between the surfaces is compatible with integral numbers of layers of water molecules.**

The hydrophobic effect: a reappraisal. P. L. Privalov & S. J. Gill (1989) *Pure Appl. Chem.* **61**, 1097–1104.

Hydrophobic hydrophilic phenomena in biochemical processes. A. Ben-Naim (2003) *Biophys. Chem.* **105**, 183–193.

Interfaces and the driving force of hydrophobic assembly. D. Chandler (2005) *Nature* **437**, 640–647.

Simple models for hydrophobic hydration. S. Hofinger & F. Zerbetto (2005) *Chem. Soc. Rev.* **34**, 1012–1020.

Entropy convergence in the hydration thermodynamics of n-alcohols. G. Graziano (2005) *J. Phys. Chem. B* **109**, 12160–12166.

3.3. MEMBRANES: HYDROPHOBIC BILAYERS IN AN AQUEOUS ENVIRONMENT

Membranes separate the various compartments of cells, and they also provide very nonpolar environments for specific proteins that are situated there. Membranes arise from the hydrophobic interaction between various lipids (Figure 3-7), which are **amphiphilic** molecules with a polar head group and very nonpolar tails made from fatty acids (Figure 3-8). With amphiphilic molecules, a useful concept is the **hydrophobic moment**. It is exactly analogous to the dipole moment of electrical charge (Section 2.2.B.1) but represents a vector from the hydrophilic to the hydrophobic parts of a molecule.

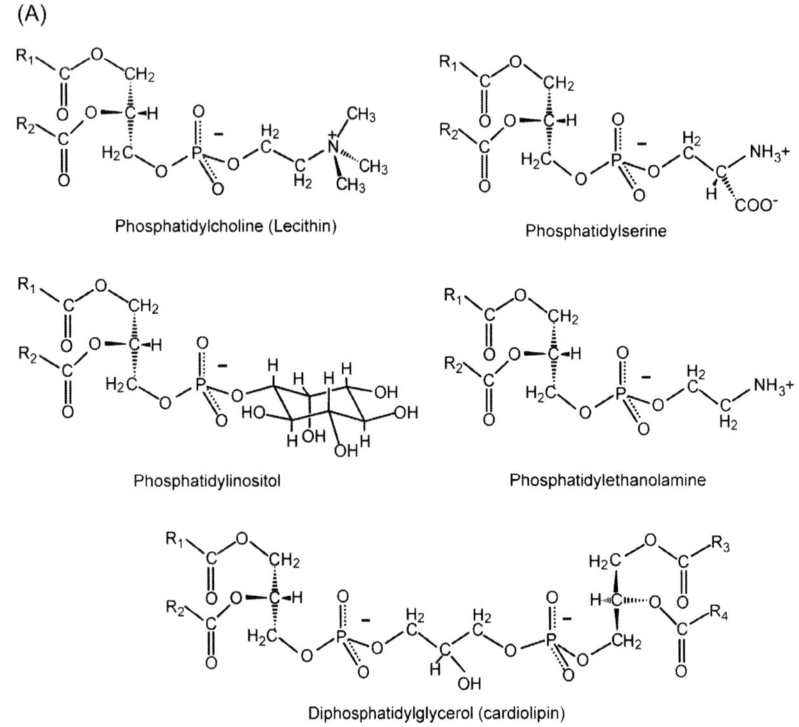

(A)

Phosphatidylcholine (Lecithin)

Phosphatidylserine

Phosphatidylinositol

Phosphatidylethanolamine

Diphosphatidylglycerol (cardiolipin)

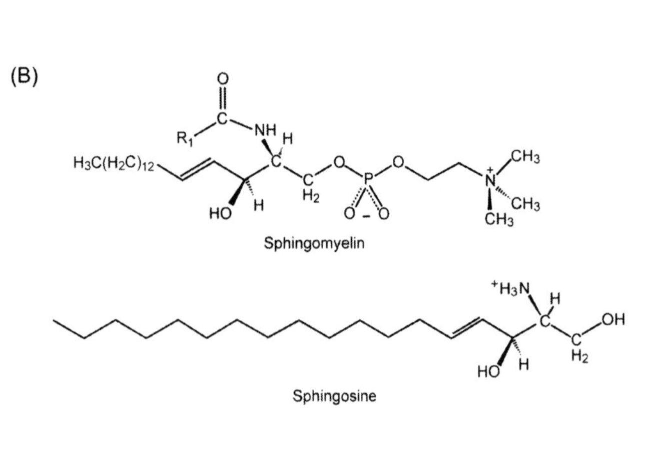

(B)

Sphingomyelin

Sphingosine

Figure 3-7. The structures of the lipids that are most common in membranes. The R_1 and R_2 moieties are fatty acids. (A) The phospholipids phosphatidylcholine (also known as lecithin), phosphatidylserine, phosphatidylinositol, phosphatidylethanolamine and diphosphatidylglycerol (also known as cardiolipin). The stereochemistry of the glycerol moiety is always that indicated. (B) The structures of sphingomyelin and sphingosine, which are the basis of glycolipids. Sphingomyelin has a fatty acid attached to the amino group of sphingosine, and phosphorylcholine attached to the hydroxyl group. Glycolipids have sugars in place of the phosphorylcholine.

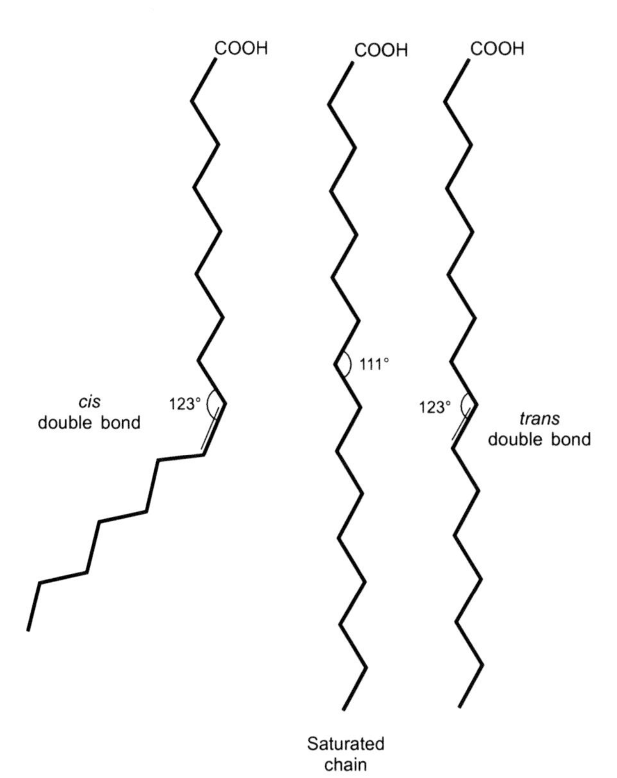

Figure 3-8. The structure of the common, saturated fatty acid palmitic acid, with 16 C atoms (*center*), and the effects of double bonds on the structure. A *trans* double bond (*right*) causes little change in structure of a saturated chain, but a *cis* double bond (*left*) produces a bend. The *trans* double bond is rare in naturally occurring fatty acids.

Most natural fatty acids contain a terminal carboxylic acid group and even numbers (14–20) of aliphatic C atoms in straight chains, because they are biosynthesized by the joining of units with two C atoms. The aliphatic chains may contain up to six nonconjugated double bonds, which are normally *cis*, having the two parts of the chain on the same side of the double bond; each pair of double bonds is generally separated by two single bonds and a single methylene group. Some of the more abundant fatty acids used in animal membranes are listed in Table 3-3. The nomenclature of fatty acids is as complex as their structures are diverse. The C atoms are numbered starting from the terminal carboxyl group, using either numbers or Greek letters; for example, the second and third C atoms can be designated as α and β, respectively. Generally, the number of C atoms in the chain is followed by a colon, the number of double bonds and their positions. The positions of the double bonds are set by the number of C atoms from the last double bond to the terminal methyl group (often designated ω). For example, linoleic acid has 18 C atoms and two double bonds, between C atoms 12–13 and 9–10, and may be written as either 18:2(n–6) or 18:2ω6 (Table 3-3). Alternatively, the position of a double bond can be represented by the symbol Δ followed by a superscript number of the first C atom of the double bond. For example, *cis*-Δ^9 means that there is a *cis* double bond between C atoms 9 and 10.

The structures of the fatty acids, their lengths and especially the number of double bonds, greatly affect their physical properties, which are reflected most simply in their melting temperatures. A *cis* double bond produces a marked kink in the fatty acid molecule (Figure 3-8), which disrupts the packing between the normally nearly straight saturated chains and lowers the melting temperature. For example, the melting point of stearic acid (with 18 C atoms and no double bonds) is about 70° C, whereas that of oleic acid (with one *cis* double bond) is only 13° C. The presence of further double bonds decreases the melting temperature further. The effect of chain length is illustrated by the 6.5° lower melting temperature of palmitic acid (with 16 C atoms) than stearic acid (with 18).

Table 3-3. Some common fatty acids

Fatty acid	Chemical name	ω-formula	Δ-formula	Melting temperature (°C)	Solubility (μM)
Saturated					
Lauric	Dodecanoic acid		12.0	44.2	11.5
Myristic	Tetradecanoic acid		14.0	54.4	0.79
Palmitic	Hexadecanoic acid		16.0	62.9	0.12
Stearic	Octadecanoic acid		18.0	69.6	0.02
Arachidic	Eicosanoic acid		20.0	75.4	0.0003
Behenic	Docosanoic acid		22.0	80.0	
Lignoceric	Tetracosanoic		24.0	84.2	
Monounsaturated					
Myristoleic	*cis*-Tetradecenoic acid	c14:1ω5	14.1 9C		2
Palmitoleic	*cis*-9-Hexadecenoic acid	c16:1ω7	16.1 9C	0.5	0.3
Oleic	*cis*-9-Octadecenoic acid	c18:1ω9	18.1 9C	α: 13.4 β: 16.3	0.05
Elaidic	*trans*-9-Octadecenoic acid	t18:1ω9	18.1 9T	46.5	
cis-Vaccenic	*cis*-11-Octadenenoic acid	c18:1ω7	18.1 11C	14.5	
Petroselinic	*cis*-6-Octadecenoic acid	c18:1ω12	18.1 6C	30	
Erucic	*cis*-13-Docosenoic acid	c22:1ω9	22.1 13C	34.7	0.007
Polyunsaturated					
Linoleic	*cis*-9,12-Octadecadienoic acid	c18:2ω6		−5	
γ-Linolenic	*cis*-6,9,12-Octadecatrienoic acid	c18:3ω6			
Linolenic	*cis*-9,12,15-Octadecatrienoic acid	c18:3ω3		−10	
Eicosatrienoic	*cis*-5,8,11-Eicosatrienoic acid	c20:3ω9			
Arachidonic	*cis*-5,8,11,14-Eicosatetraenoic acid	c20:4ω6		−50	
Eicosapentenoic	*cis*-5,8,11,14,17-Eicosapentaenoic acid	c20:5ω3			

Fatty acids are generally part of the lipids that make up natural membranes, which are usually either **phospholipids** or **glycolipids**. A phospholipid is constructed of two fatty acids attached to C-1 and C-2 of the three-carbon alcohol **glycerol** and a phosphate group on C-3, to which an alcohol is attached, usually serine, ethanolamine, choline, glycerol or inositol. The glycerol moiety always has the stereochemistry illustrated in Figure 3-7. Glycolipids are sugar-containing lipids derived from sphingosine, but they have one or more sugars attached in place of the phosphorylcholine of sphingomyelin.

When dissolved in water, lipids spontaneously aggregate via their nonpolar tails in order to minimize their contact with water. What type of structure they adopt depends upon the conditions and, especially, upon the relative structures of the polar and nonpolar parts of the lipids (Figure 3-9). In each case, the primary driving force is the hydrophobic interaction described in Section 3.2, but there

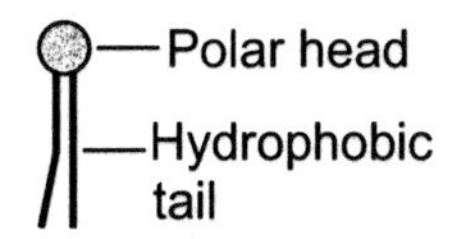

(A) Lipid molecule

(B) Micelle

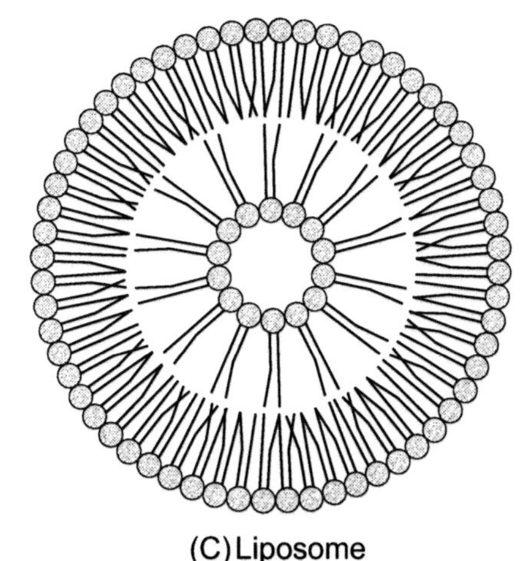

(C) Liposome

(D) Bilayer

Figure 3-9. Schematic illustrations of the structures formed by amphiphiles in aqueous solution. (A) Schematic diagram of a single amphipathic lipid molecule. (B) Lipids with single long nonpolar tails tend to form micelles, in which the molecules aggregate into spheres with the tails in the interior, not in contact with the water. (C) Curved bilayers tend to result when the two layers have different structures, and they can close up to form spherical liposomes. They have an aqueous interior. (D) Lipids with two long nonpolar tails, such as phospholipids, have similar cross-sections for their polar head groups and the tails, so they form planar monolayers and bilayers. Monolayers are generated at nonpolar surfaces, such as the interface between water and air, where the polar head groups are in contact with the water. Bilayers are formed in aqueous solution, to minimize the contact of the nonpolar tails with water.

can also be polar interactions between the neighboring head groups. Amphiphilic molecules like lipids can form a **monolayer** at the surface of an aqueous solution, with their polar groups in contact with the water and their nonpolar parts extending into the air. They can, especially if the nonpolar tail is thinner in structure than the head group, as in individual fatty acids, aggregate spontaneously into very stable spherical **micelles**, where the nonpolar tails occupy the center, and exclude water, and the polar head groups are on the surface, interacting favorably with the water (Figure 3-9-B). Micelle formation is a cooperative process, because an assembly of just a few amphiphiles cannot shield their tails from contact with water. Consequently, dilute aqueous solutions of amphiphiles do not form micelles until their concentration surpasses a certain **critical micelle concentration (cmc)**.

When the head groups and nonpolar tails of lipids have similar cross-sections, as occurs in most phospholipids and glycolipids, they tend to pack side-by-side to generate a flat sheet with one polar face and one nonpolar face. In water, the hydrophobic surfaces aggregate to form a **lipid bilayer** (Figure 3-9-C). This is the favored structure for most phospholipids in water, rather than micelles. If different lipid molecules make up the two layers, they tend to associate into curved layers. A lipid bilayer can close to become a **lipid vesicle**, or **liposome**, which encloses an aqueous interior (Figure 3-9-C). The surface of the bilayer accessible to the aqueous solvent depends upon the nature of the head groups.

At low temperatures, a pure phospholipid containing a single species of saturated fatty acid forms relatively rigid, quasi-crystalline bilayers, which 'melt' at a characteristic temperature to a more liquid crystalline phase (Figure 3-10). The longer the nonpolar fatty acid chains and the fewer the double bonds, the higher the melting temperature. The more heterogeneous the fatty acid chains and, more especially, the greater the degree of unsaturation, the lower and less well-defined the transition temperature.

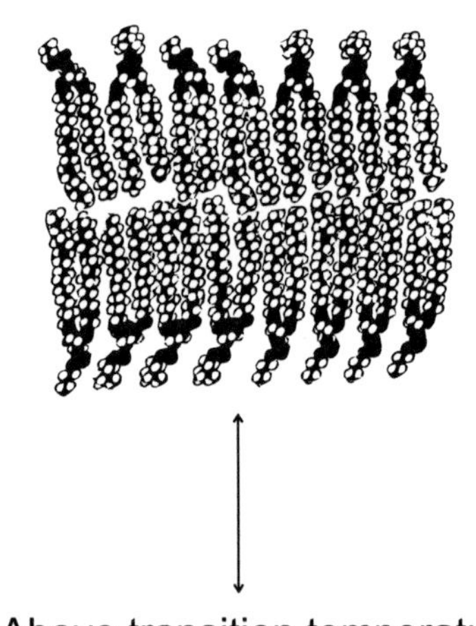

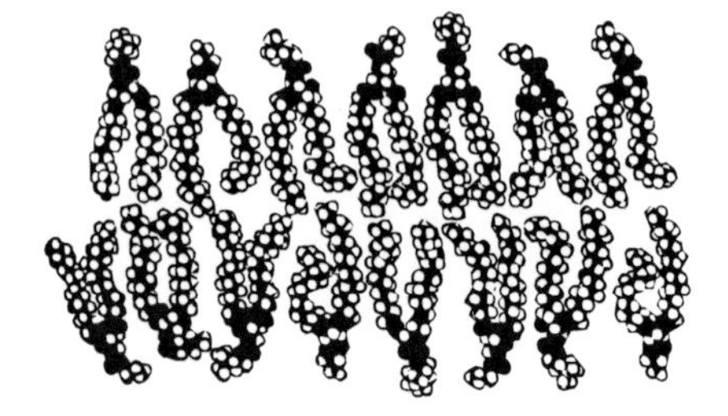

Figure 3-10. Schematic illustration of the 'melting' of a lipid bilayer composed of phosphatidylcholine and phosphatidylethanolamine as the temperature is increased. At low temperatures (*top*), the lipid molecules are in an orderly array, a gel-like solid. Above the transition temperature (*bottom*), the lipid molecules and their nonpolar tails are highly mobile in the plane of the bilayer. Adapted from R. N. Robertson.

Lipid bilayers provide a nonpolar interior that is permeable only to nonpolar molecules and impermeable to most polar molecules and ions such as Na^+, K^+, and Cl^-. Water is an apparent exception, in that it can readily diffuse across the bilayer; this is probably due to its small size, high concentration and absence of a net charge. Lipid-soluble nonpolar substances, such as triacylglycerol and nonionized organic acids, readily diffuse into the bilayer and remain within its nonpolar interior. Nonpolar molecules partition into the membrane bilayer similarly as they do with nonpolar liquids, but those with polar groups as well (such as peptides) bind only to the interface between the head groups and the nonpolar layer. The main difference with partitioning into nonpolar liquids is that aromatic groups bind more tightly to membranes than aliphatic molecules do, indicating that there may be specific interactions with them.

The lipids are free to move within the bilayer by lateral diffusion, with a typical speed of 1–2 μm/s, and the interior has a viscosity comparable to that of castor oil and solvent properties similar to those of octanol. This means that a lipid molecule can diffuse from one end of a bacterium to the other within one second. The lipids very rarely leave the bilayer, and they only infrequently reorient from one surface to the other, with a spontaneous half-time of about day, because to do so requires the polar head group to traverse the nonpolar membrane interior.

Natural membranes are composed of mixtures of a great variety of phospholipids, glycolipids and sphingolipids that vary in both their head groups and fatty acid tails. The fatty acid compositions of natural membranes vary widely, but there is always a sufficient number of double bonds for the membrane to have a fluid lipid phase at the physiological temperature. Organisms that grow at low temperatures have a greater proportion of unsaturated fatty acids in their membranes. Most natural membranes undergo this transition in the range of 10–40° C.

Cholesterol is also an important component of many natural membranes, in spite of being a steroid rather than a lipid; in animals it is the primary regulator of the fluidity of the membrane. It has four fused rings and an eight-member branched hydrocarbon chain that make it a compact, rigid, hydrophobic molecule:

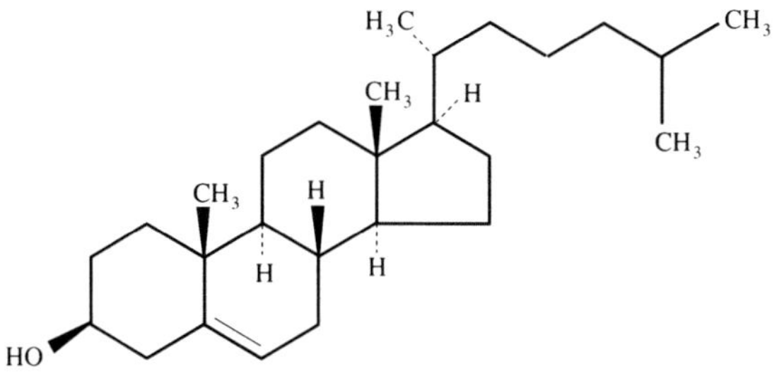

(3.4)

It is also amphiphilic because of the polar hydroxyl group. Although it does not form a bilayer by itself, its presence in membranes produces a more condensed and impermeable bilayer; the fluidity of the bilayer and its hydrophobicity are decreased. It also broadens the order–disorder transition and totally abolishes it at high concentrations. Its rigid steroid ring system apparently fits between the fatty acid side-chains of the lipids, but its different shape interferes with their motions and inhibits their crystallization. Its hydroxyl group is believed to interact with the polar head groups of the lipids.

In most natural membranes, the two layers have different lipid compositions. In erythrocytes, for example, the external monolayer of the membrane bilayer contains mostly neutral phospholipids, such as phosphatidylcholine and sphingomyelin, while the internal monolayer contains primarily phosphatidylserine and phosphatidylethanolamine. Glycolipids are almost always on the external side of a natural membrane. This asymmetric distribution of lipids arises from their biosynthesis by growth of pre-existing membranes and is maintained by metabolic processes; when they are disrupted, the bilayer slowly reverts to the more stable symmetric lipid distribution across the bilayer.

The composition of each layer is apparently not uniform; instead the various lipids tend to segregate to form 'domains'. For example, mixtures of cholesterol and phospholipid in monolayers and bilayers are subject to liquid–liquid immiscibility, and some mixtures exhibit several such types of immiscibilities. This unusual property has led to the proposal of 'condensed complexes', which result from an exothermic, reversible interaction between cholesterol and phospholipids. The complexes are sometimes concentrated in a separate liquid phase. The phase separation into a complex-rich phase depends on the membrane composition and intensive variables such as the temperature.

The heterogeneity and fluid nature of membranes means that their atomic-scale structures are illustrated best by computer simulations (Figure 3-11). Membranes commonly fuse and divide during many processes in living cells, and how they do this is a topic of major interest. Model membranes will do so spontaneously in aqueous solution, but in living cells their fusion involves interactions between the proteins embedded within the membranes.

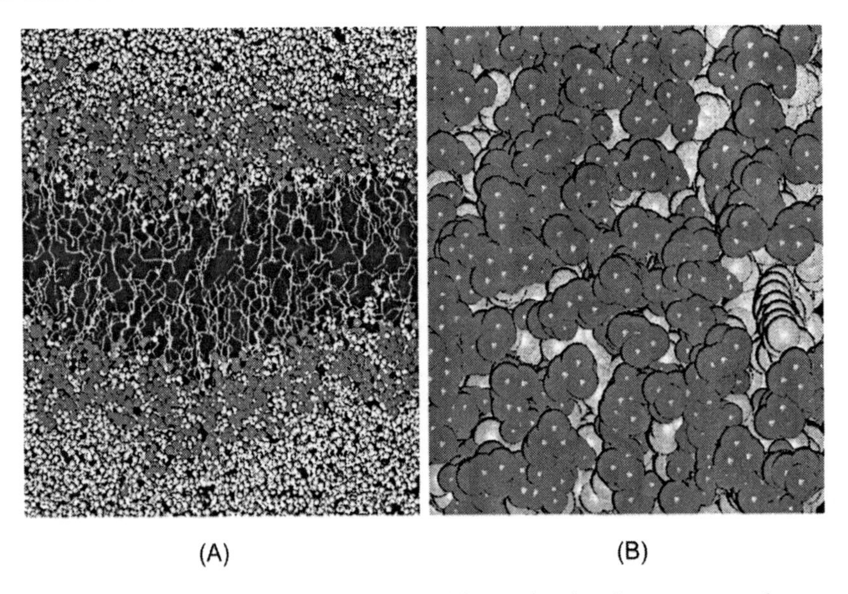

Figure 3-11. Membrane structure in water as revealed by molecular dynamics simulations. (A) Details of a fluid-phase, fully hydrated bilayer of dimyristoylphosphatidylcholine. The phospholipid head groups are dark, the hydrocarbon tails light; water molecules are white. The sizes of the atoms are reduced from their van der Waals dimensions to permit seeing the interior of the structure and the interpenetration of water with the phospholipid head groups. (B) A view of the same membrane but from the top, looking down onto the membrane surface. The water molecules have been removed, and the atoms are shown with their full van der Waals radii. The gaps between the head groups through which the hydrocarbon tails are visible would be filled with water.

Many proteins reside specifically within the interiors of natural membranes, where they are usually inserted during their biosynthesis. They can account for 30–80% of the weight of the membrane. Yet they are free to diffuse and have relatively small effects on the properties of the bilayer; they affect primarily the lipid molecules in close contact with them.

The hydrophobic moment and its use in the classification of amphiphilic structures. D. A. Phoenix & F. Harris (2002) *Mol. Membr. Biol.* **19**, 1–10.

Liquid–liquid immiscibility in membranes. H. M. McConnell & M. Vrljic (2003) *Ann. Rev. Biophys. Biomolec. Structure* **32**, 469–492.

New insights into water–phospholipid model membrane interactions. J. Milhaud (2004) *Biochim. Biophys. Acta* **1663**, 19–51.

Lipid bilayers: thermodynamics, structure, fluctuations, and interactions. S. Tristram-Nagle & J. F. Nagle (2004) *Chem. Phys. Lipids* **127**, 3–14.

Rafts, little caves and large potholes: how lipid structure interacts with membrane proteins to create functionally diverse membrane environments. R. Morris *et al.* (2004) *Subcell. Biochem.* **37**, 35–118.

Roles of bilayer material properties in function and distribution of membrane proteins. T. J. McIntosh & S. A. Simon (2006) *Ann. Rev. Biophys. Biomolec. Structure* **35**, 177–198.

Partial molecular volumes of lipids and cholesterol. A. I. Greenwood *et al.* (2006) *Chem. Phys. Lipids* **143**, 1–10.

3.3.A. Detergents

Membranes must often be disrupted in order to isolate components embedded within them, especially the proteins. This is usually accomplished with **detergents, which are amphiphilic molecules that can replace the lipids that make up the membrane but do not form membrane-like structures.** Detergents are relatively soluble in aqueous solution, but they form micelles above their critical micelle concentration.

Detergents are usually classified according to whether or not their head groups are strongly ionic, either negatively or positively charged, such as SDS and cetylpyridinium chloride (Figure 3-12). Others have head groups that are not charged but have multiple hydrogen-bonding groups, such as octanoyl-*N*-methylglucamide (MEGA-8), Thesit, *n*-dodecyl-β-maltoside and *n*-octyl-β-glucoside. Triton X-100 and X-114 have long polyoxyether chains and aromatic rings. CHAPS and CHAPSO are zwitterions, with both positive and negative charges.

Detergents differ in their ability to solublize proteins and lipids from membranes (Table 3-4) but there are no rules as to which detergent is best for any particular membrane or protein. The amount of detergent used is generally in the range of 0.5–2.0% weight per volume (w/v), i.e. 0.5–2.0 g/100 ml. Proteins are usually extracted from membranes most successfully when the detergent present is 1–3 times the weight of protein present. Detergents such as SDS tend to denature proteins, whereas CHAPS, CHAPSO and octylgucoside are less denaturing and more likely to maintain a membrane protein in a functionally active state. Increasing the concentration of a detergent gradually extracts one protein/lipid after another. The spatial heterogeneity of membranes results in different areas having different susceptibilities to detergents; a fraction that is detergent-resistant tends to have certain proteins in it.

The excess detergent can be removed by dialysis, gel filtration, or hydrophobic chromatography. It is much easier to remove excess lipids when they are monomeric than when in large micelles, so the critical micelle concentration of a detergent is important. The physical states of lipids and proteins solubilized with detergents in this way are generally not well-characterized.

Detergents for the stabilization and crystallization of membrane proteins. G. G. Prive (2007) *Methods* **41**, 388–397.

Two distinct mechanisms of vesicle-to-micelle and micelle-to-vesicle transition are mediated by the packing parameter of phospholipid–detergent systems. M. C. Stuart & E. J. Boekema (2007) *Biochim. Biophys. Acta* **1768**, 2681–2689.

Triton X-100 partitioning into sphingomyelin bilayers at subsolubilizing detergent concentrations. Effect of lipid phase and a comparison with dipalmitoylphosphatidylcholine. C. Arnulphi *et al.* (2007) *Biophys. J.* **93**, 3504–3514.

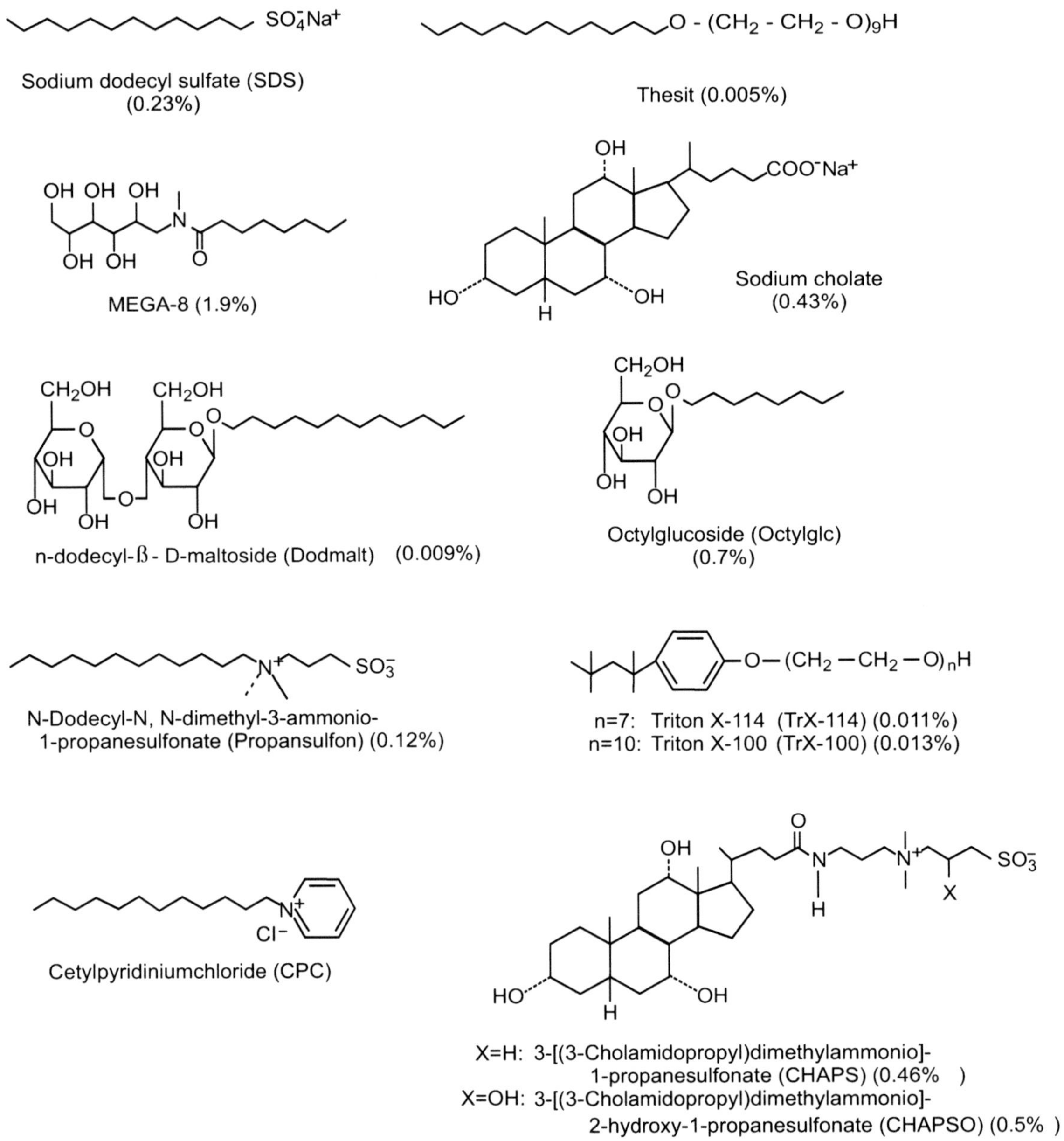

$SO_4^-Na^+$

Sodium dodecyl sulfate (SDS) (0.23%)

$O - (CH_2 - CH_2 - O)_9H$

Thesit (0.005%)

MEGA-8 (1.9%)

COO^-Na^+

Sodium cholate (0.43%)

n-dodecyl-ß- D-maltoside (Dodmalt) (0.009%)

Octylglucoside (Octylglc) (0.7%)

SO_3^-

N-Dodecyl-N, N-dimethyl-3-ammonio-1-propanesulfonate (Propansulfon) (0.12%)

$O-(CH_2-CH_2-O)_nH$

n=7: Triton X-114 (TrX-114) (0.011%)
n=10: Triton X-100 (TrX-100) (0.013%)

Cl^-

Cetylpyridiniumchloride (CPC)

X=H: 3-[(3-Cholamidopropyl)dimethylammonio]-1-propanesulfonate (CHAPS) (0.46%)
X=OH: 3-[(3-Cholamidopropyl)dimethylammonio]-2-hydroxy-1-propanesulfonate (CHAPSO) (0.5%)

Figure 3-12. The structures of some commonly used detergents and their abbreviations. The critical micelle concentration of each is given in units of weight of detergent per volume of aqueous solution.

Table 3-4. Extraction and solubilization of proteins and lipids from membranes by various detergents

Detergent	Solubilized material (mg/ml)	
	Proteins	Lipids
Cetylpyridinium chloride (CPC)	9.1	1.3
Triton X-100	6.2	1.5
Thesit	5.9	2.0
Propane sulfonate	5.1	4.4
SDS	4.6	5.4
Octyl glucoside	4.4	4.0
Dodecyl maltoside	3.9	5.8
Triton X-114	2.9	2.8
Sodium cholate	2.5	4.9
CHAPSO	2.5	6.5
CHAPS	2.5	7.7

After treatment of the membranes with the detergents of Figure 3-12, the soluble fraction was dialyzed to remove the free detergents and centrifuged to remove insoluble material. The lipids and proteins present in the soluble material were measured. Data from P. Banerjee (1999), in *Encyclopedia of Molecular Biology* (T. E. Creighton, ed.), Wiley-Interscience, NY, p. 664.

3.4. IONIZATION

Water spontaneously dissociates into one hydrogen ion (proton) and one hydroxide ion:

$$H_2O \leftrightarrow H^+ + OH^- \tag{3.5}$$

At 25° C:

$$K_{eq} = [H^+]\,[OH^-]/[H_2O] = 10^{-14}\ \text{M} \tag{3.6}$$

The value of the equilibrium constant changes somewhat at different temperatures: it is 0.12, 2.9 and 5.4 ($\times\ 10^{-14}$ M) at 0°, 40° and 100° C, respectively. **In pure water, the concentrations of H^+ and OH^- will be equal and about 10^{-7} M, i.e. 0.1 μM.** The pH is said to be neutral, close to 7.*

The pH is defined as:

$$pH = -\log_{10}\,[H^+]\ \text{M}^{-1} \tag{3.7}$$

* *This discussion assumes that all concentrations are sufficiently small so that ideality applies. If not, all the concentrations should be corrected with the appropriate activity coefficients.*

The pH scale is useful in water over about the range 0–14, when at 25° C $[H^+]$ varies from 1 M to 10^{-14} M, while $[OH^-]$ varies conversely from 10^{-14} M to 1 M.

The H^+ and OH^- ions are largely hydrated, being associated reversibly and rapidly with one or more water molecules. The average lifetime of a proton attached to a water molecule at neutral pH and room temperature is approximately 0.4 ms.

Other polar molecules will also release or accept protons in water. Using the Brønsted definition, an **acid** is a substance (AH) that can donate a proton:

$$AH \leftrightarrow A^- + H^+ \tag{3.8}$$

while a **base** (B) can combine with one:

$$B + H^+ \leftrightarrow BH^+ \tag{3.9}$$

In this sense, A^- can be considered to be the **conjugate base** of the acid AH, while BH^+ is the **conjugate acid** of base B. Note that the species A and B might have other charged groups, so species A and B need not have the overall charges indicated above. In each case, however, the group that releases or accepts the proton and acquires a net charge is said to **ionize**. The tendency of each acid or base to ionize will be given by the equilibrium constant for the above reactions. Considering the acidic species:

$$K_a = [H^+]\,[A^-]/[HA] \tag{3.10}$$

This defines the pK_a of the acidic group, comparable to the definition of the pH: $\mathbf{-log_{10}\, K_a = pK_a}$. Rearranging Equation 3.10 gives:

$$[H^+] = K_a\,[HA]/[A^-] \tag{3.11}$$

Using the definition of pH (Equation 3.7), Equation 3.11 becomes:

$$pH = pK_a - \log_{10}\,[A^-]/[HA] \tag{3.12}$$

This is the **Henderson–Hasselbalch equation** that is fundamental for all considerations of pH and ionization.

The degree of ionization of an acid, i, is $10^{-pK_{ai}} / \left(10^{-pK_{ai}} + 10^{-pH}\right)$, and the degree of ionization of a base, j, is $10^{-pH} / \left(10^{-pK_{ai}} + 10^{-pH}\right)$. When the pH is the same as the pK_a, the acid HA or the base B is half-ionized, for example $[A^-] = [HA]$. When the pH is one pH unit greater than the pK_a, $[A^-] = 11 \times [HA]$, and at one pH unit lower than the pK_a, $[A^-] = 0.09 \times [HA]$. Consequently, complete ionization of a single group in isolation requires a change in pH of more than 2 pH units. The variation with pH of the ionization of a single group in isolation is described by its **ionization curve** (Figure 3-13). All simple acids with a single acidic group in isolation follow the ionization curve of Figure 3-13; they differ only in their pK_a values, where the ionization curve is centered on the pH scale. The intrinsic pK_a values of groups in biological macromolecules are given in Table 3-5.

Table 3-5. The pK_a values for amino acids in peptides and bases in nucleosides

Amino acid residues in peptides	pK_a	Bases in nucleosides	pK_a
α-NH_2	7.8	Adenine (N^1)	3.52
α-COOH	3.6	Cytosine (N^3)	4.17
Asp ($C^{\gamma}OOH$)	4.0	Guanine (N^1)	9.42
Glu ($C^{\delta}OOH$)	4.5	Guanine (N^7)	3.3
His (imidazole)	6.4	Uracil (N^3)	9.38
Lys (ε-NH_2)	10.4	Thymine (N^3)	9.93
Arg (guanidine)	≈12		
Tyr ($O^{\eta}H$)	9.7		
Cys ($S^{\gamma}H$)	9.1		

Data from A. R. Fersht (1985) *Enzyme Structure and Mechanism*, 2nd edn, Freeman, NY, Table 5.1, and G. M. Blackburn (1996) in *Nucleic Acids in Chemistry and Biology: DNA and RNA structure*, 2nd edn (G. M. Blackburn and M. J. Gait, eds), Oxford University Press, Oxford, Table 6-TAB.

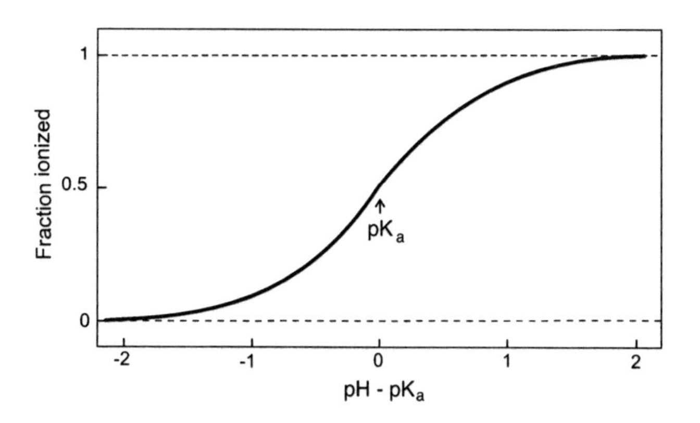

Figure 3-13. An ionization curve for a single group in isolation. The fraction of the basic form (A^-) of a monobasic acid (HA) is depicted as a function of the pH. The pH is expressed relative to the pK_a of the group.

The titration curve of one ionizing group can be affected by the ionization of other groups nearby. With multiple groups ionizing, very different ionization curves can result if they are interdependent. Interactions between ionized groups invariably involve changes in their pK_a values, and the ionization of one group can alter the ionization of another nearby. **Favorable electrostatic interactions involving the ionized form increase the tendency of any group to ionize, whereas repulsions have the opposite effect.** Other factors also influence the pK_a value, however, such as the accessibility to the solvent and its polarity. For example, adding dioxane to **decrease the polarity of the solvent inhibits ionization of accessible amino and carboxyl groups** (Figure 3-14). The apparent pK_a values of amino groups decrease, whereas those of carboxyl groups increase. In contrast, the largely inaccessible amino group of Tris is not affected. **Ionization also becomes less favorable with increasing bulkiness of the surrounding aliphatic groups** (Table 3-6).

Table 3-6. Steric effects on the ionization of carboxyl groups

Model compound	pK_a value
H_3C-CO_2H	5.55
$CH_3-C(CH_3)_2-CH(CH_3)-CO_2H$	6.25
$CH_3-CH_2-C(CH_2CH_3)(CH_2CH_3)-CO_2H$	6.44
$CH_3-C(CH_3)_2-C(CH_3)_2-CO_2H$	6.71
$H_3C-C(CH_3)_2-CH_2-C(CH_3)(C(CH_3)_2CH_3)-CO_2H$	6.97

The pK_a values were measured in equal volumes of methanol and water at 40° C by G. S. Hammond & D. H. Hogle (1955) *J. Am. Chem. Soc.* **77**, 338–340.

Protonation reactions occur at rates approximately those expected if they are limited by the rates at which the reactants come into contact by diffusion, with second-order rate constants of roughly 10^{10} s^{-1} M^{-1}. It is then generally assumed that they occur very rapidly in solution, but this need not be the case if the concentration of H^+ is very low. For example, at pH 10, $[H^+] = 10^{-10}$ M, so the apparent rate constant for protonation of a molecule will be only 1 s^{-1}.

pH and Buffer Theory: a new approach. H. Rilbe (1996) John Wiley, Chichester.

Autoionization in liquid water. P. L. Geissler *et al.* (2001) *Science* **291**, 2121–2124.

Determination of microscopic acid–base parameters from NMR–pH titrations. Z. Szakacs *et al.* (2004) *Anal. Bioanal. Chem.* **378**, 1428–1448.

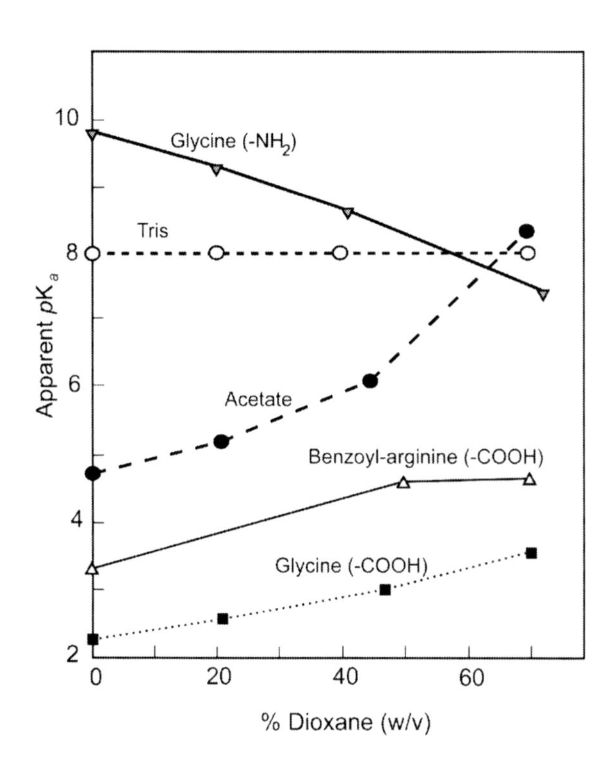

Figure 3-14. Effect of a nonaqueous environment on the pK_a values of amino and carboxyl groups. The apparent pK_a was measured at various concentrations of dioxane. Data from A. Fersht (1977) *Enzyme Structure and Mechanism*, W. H. Freeman, Reading.

3.4.A. Measuring the pH

Every laboratory possesses a pH meter that is routinely used to measure and adjust the pH values of solutions. Yet obtaining an accurate measure is not as simple as is commonly assumed, and considerable care is required. Most pH meters use a glass electrode, which in the solution to be measured creates an electrical cell in which the potential of this cell depends upon the pH of the solution. A layer of thin glass separates the solution being measured from one containing one of the electrodes of the cell. **The potential across it depends upon the pH of the solution because the thin glass of the electrode is selectively permeable only to protons**. The bridge to the other electrode contains concentrated KCl, and its junction potential should be negligible, so the measured potential should be proportional to the concentration of protons, $[H^+]$, in the solution being measured. Glass membranes may, however, be slightly permeable to Na^+ ions, so their presence in high concentrations may contribute to the potential, especially when the concentration of protons is low, at alkaline pH. The $[H^+]$ in alkaline solutions will then be overestimated and the pH underestimated.

pH meters require calibration, and standard buffers are available commercially for this. It is especially important to control the temperature; do not assume that a pH meter calibrated at one temperature will give an accurate reading at another. Meters will usually give accurate results only over a limited pH range, so the calibration should be made at a pH close to that to be measured. Be aware that all buffers and solutions with a high pH will absorb carbon dioxide from the air and become more acidic with time.

Specific anion effects on glass electrode pH measurements of buffer solutions: bulk and surface phenomena. A. Salis *et al.* (2006) *J. Phys. Chem. B* **110**, 2949–2956.

3.4.B. Buffers

The concentrations of H^+ and OH^- ions are very small in the neutral pH region around pH 7, so any process that generates or consumes significant amounts of either can cause the pH of the solution to fluctuate wildly. For example, a biochemical reaction that occurred at pH 7 but produced 1 mM H^+ would decrease the pH to 3 in pure water. A buffer prevents this and is obligatory in every biochemical system. The principle of a buffer is very simple: a buffer consists of high concentrations of a base (A) and its conjugate acid (HA) in the correct proportions to produce the desired pH value, and at concentrations much higher than those of the H^+ and OH^- ions present or likely to be produced during any reaction. Those H^+ and OH^- ions that are generated simply displace the equilibrium of Equation 3.10 very slightly, and the pH changes very little. Equation 3.12 can be rearranged to:

$$pH = pK_a - \log_{10} [A^-] + \log_{10} [HA] \tag{3.13}$$

This demonstrates that the pH will remain constant only if both the $\log_{10} [A^-]$ and the $\log_{10} [HA]$ terms remain reasonably constant. This requires that both terms be substantial, so both HA and A^- must be present at substantial concentrations. Consequently, **a buffer must have a pK_a value close to the value of the desired pH**. The pK_a values of some commonly used buffers are given in Table 3-7.

The reactive components of a biochemical reaction mixture might have sufficient buffering capacity themselves. If not, a buffer reagent must be added. Unfortunately, there are relatively few buffers with pK_a values close to neutrality, which would be most useful in physiological studies. Most of the useful buffers are amines with electron-attracting substituents that make the lone-pair of electrons on the N atom less available and so lower the atom's pK_a. Examples include Tris and 2-morpholinoethanesulfonic acid:

$$(HO-CH_2)_3C-NH_2 \qquad O(CH_2-CH_2)_2N-CH_2-CH_2-SO_3H \tag{3.14}$$

Tris 2-Morpholinoethanesulfonic acid

which have pK_a values of 8.1 and 6.2, respectively. Buffers that rely on amino groups have the disadvantages that their pK_a values change with the temperature and that primary amino groups can react with some chemical reagents. Phosphate buffers are often used because interconverting $H_2PO_4^-$ and HPO_4^{2-} occurs with a pK_a close to 7, but phosphate is involved in many biochemical reactions, so it is then not a neutral component of such a mixture. Its pK_a is also sensitive to the salt concentration (Section 3.4.B.1).

Buffer solutions are usually specified in a form such as 0.2 M sodium acetate buffer, pH 4.8. The concentration should refer to the sum of the two forms of the buffering molecule, in this case acetate ion (CH_3-CO_2^-) and acetic acid (CH_3-CO_2H). This buffer can be prepared correctly in any of three different ways:

(i) mixing the appropriate amounts of sodium acetate and acetic acid as calculated from the pK_a and the desired pH (Equation 3.12);

(ii) adding 0.2 M sodium acetate to 0.2 M acetic acid until pH 4.8 is reached;

(iii) adding concentrated NaOH to 0.2 M acetic acid to reach pH 4.8.

Table 3-7. Some common buffers used in molecular biology

Buffer	pK_a
Histidine	1.82, 5.98, 9.17
Phosphate	1.96, 6.7, 12.3
Formate	3.75
Barbiturate	3.98
Acetate	4.8
Pyridine	5.23
Bis-tris (bis-(2-hydroxyethyl)imino-tris-(hydroxymethyl)-methane)	6.46
Pipes (1,4-piperazinebis-(ethanesulfonic acid))	6.8
Imidazole	7.0
Bes (*N,N*-bis-(2-hydroxyethyl)-2-aminoethane-sulfonic acid)	7.15
Mops (2-(*N*-morpholino)propane-sulfonic acid)	7.2
Hepes (*N*-2-hydroxyethyl-piperazine-*N′*-ethanesulfonic acid)	7.55
Tris-hydroxymethyl-amino-methane	8.1
Taps (*N*-tris(hydroxymethyl)methyl-2-aminopropane sulfonic acid)	8.4
Borate	9.39
Ethanolamine	9.44
Caps (3-cyclohexylamino-1-propane-sulfonic acid)	10.4
Methylamine	10.64
Triethylamine	10.72
Dimethylamine	10.75

It would not, however, be appropriate to add strong HCl to 0.2 M sodium acetate to reach pH 4.8, as this would also produce NaCl.

The pK_a values of buffers tend to vary with temperature to varying extents, so it is important to control the temperature when regulating the pH. The variation of the pK_a with temperature depends upon the change in the enthalpy (Section 1.3) upon titrating the buffering group. Dissociations of carboxylic acids and phosphoric acid have small enthalpy changes, so their pK_a values do not vary much with temperature. Amino groups, on the other hand, release heat when they bind protons, so increasing the temperature inhibits their protonation and lowers the pK_a. Consequently, the pH of a buffer that uses an amino group decreases upon heating. The effect can be quite large: a decrease of 0.028 pH units per °C.

The pH of a buffered solution should not vary as it is diluted or concentrated, so long as the concentration of the buffer remains sufficient, because the ratio of base and acid forms remains constant. On the other hand, the pK_a of the buffer can change as the ionic strength of the solution changes. This effect is minimal with simple buffers that exist with either 0 or 1 charges, but with a greater number of charges, either negative or positive, there will be electrostatic interactions between them, which will be screened to varying extents by the other ions in the solution. The pK_a will then vary with the salt concentration.

DNA and buffers: are there any noninteracting, neutral pH buffers? N. C. Stellwagen *et al.* (2000) *Anal. Biochem.* **287**, 167–175.

Semi-mechanistic partial buffer approach to modeling pH, the buffer properties, and the distribution of ionic species in complex solutions. D. P. Dougherty *et al.* (2006) *J. Agric. Food Chem.* **54**, 6021–6029.

Similarity of salt influences on the pH of buffers, polyelectrolytes, and proteins. A. E. Voinescu *et al.* (2006) *J. Phys. Chem. B* **110**, 8870–8876.

1. Phosphate Buffers

Orthophosphoric acid (O=P(–OH)$_3$, often abbreviated P$_i$) has three hydroxyl groups that are identical, yet they titrate with very different pK_a values of about 2, 7 and 12. The first proton to dissociate does so from any of the three groups, at about pH 2. It is much more difficult for the second proton to dissociate because of electrostatic repulsions within the doubly charged phosphate group that would result. Consequently, its pK_a is much higher, about 7. The electrostatic interactions are even stronger with the third proton, so its pK_a is much greater, about 12. Consequently, the Henderson–Hasselbalch equation for orthophosphate is:

$$\mathrm{pH} = 2 + \log\frac{[\mathrm{H_2PO_4^-}]}{[\mathrm{H_3PO_4}]} = 7 + \log\frac{[\mathrm{HPO_4^{2-}}]}{[\mathrm{H_2PO_4^-}]} = 12 + \log\frac{[\mathrm{PO_4^{3-}}]}{[\mathrm{HPO_4^{2-}}]} \tag{3.15}$$

Although it has three widely spaced pK_a values, phosphate is not a good buffer for the intermediate pH ranges 3-6 and 8–11.

Phosphate buffers have the great advantage of being transparent to UV light of very short wavelengths, so that the absorbance of the peptide bonds of proteins at about 210 nm can be monitored in this buffer. Its main disadvantages are that it supports the growth of algae and fungi and that its pK_a values are sensitive to the ionic strength (Equation 3.27), because they depend upon electrostatic interactions within the phosphate group. In the neutral pH region, where the acid is $H_2PO_4^-$ and its conjugate base is the doubly charged HPO_4^{2-}, **a solution of 0.2 M phosphate buffer may increase its pH by more than 0.2 pH units upon a 10-fold dilution to 0.02 M**. Phosphate will also sequester and precipitate many cations, especially Ca^{2+}.

An alternative to orthophosphate is **pyrophosphate** (or diphosphate, often abbreviated PP_i):

$$HO-\overset{\overset{\displaystyle O}{\|}}{\underset{\underset{\displaystyle OH}{|}}{P}}-O-\overset{\overset{\displaystyle O}{\|}}{\underset{\underset{\displaystyle OH}{|}}{P}}-OH \tag{3.16}$$

The first proton to dissociate from either of the identical phosphate groups is analogous to that with orthophosphate, and both do so with pK_a values close to 2. The interactions within the more complex pyrophosphate group result in the subsequent ionizations having somewhat different pK_a values from orthophosphate. The third proton to dissociate is subjected to strong electrostatic interactions, and the

$$-\overset{\overset{\displaystyle O}{\|}}{\underset{\underset{\displaystyle OH}{|}}{P}}-O^- \tag{3.17}$$

group is electron-withdrawing, so its apparent pK_a is about 6.2. The

$$-\overset{\overset{\displaystyle O}{\|}}{\underset{\underset{\displaystyle O^-}{|}}{P}}-O^- \tag{3.18}$$

group is slightly electron-donating to the other phosphate group, so the final proton dissociates with a pK_a of only about 8.4, even though it results in a pyrophosphate molecule with four negative charges. The Henderson–Hasselbalch equation for pyrophosphate in the neutral pH range is therefore:

$$pH = 6.2 + \log\frac{[HP_2O_7^{3-}]}{[H_2P_2O_7^{2-}]} = 8.4 + \log\frac{[P_2O_7^{4-}]}{[HP_2O_7^{3-}]} \tag{3.19}$$

The more complex and widespread interactions that take place in the pyrophosphate molecule result in its pK_a values being even more sensitive to the ionic strength than those of orthophosphate.

New aspects of buffering with multivalent weak acids in capillary zone electrophoresis: pros and cons of the phosphate buffer. P. Gebauer & P. Bocek (2000) *Electrophoresis* **21**, 2809–2813.

---

2. *Tris Buffer*

Tris is one of the most widely used buffers. Its common name comes from its original, but now out-dated, name tris(hydroxymethyl)methylamine (Equation 3.14). The three hydroxyl groups are electron-withdrawing, so the lone pair of electrons on the N atom is much less available than on a typical primary amine, and its pK_a is reduced to 8.1 at 25° C. The pK_a value of 8.1, however, is somewhat too high for Tris to be a good buffer at neutral pH.

As with other amino groups, its pK_a varies substantially with the temperature, and it can react chemically with many reagents, although this is minimized by steric interference by the three hydroxymethyl groups.

3. *Membrane-Impermeable Good Buffers*

When working with cells and other organelles enclosed by membranes, it is best if both basic and acidic forms of a buffer are ionized, so that neither passes readily through membranes. Otherwise, for example using an acetate buffer, the acetic acid, but not acetate ions, may permeate the membranes and lower the internal pH below that of the external buffer. A number of buffers that avoided this problem were introduced by Good and colleagues, and are known as **'Good' buffers.** Many have strongly acidic groups, such as sulfonate ($-SO_3^-$), so they are ionized over most of the pH range and do not penetrate membranes. An example is 2-morpholinoethanesulfonate (Equation 3.14). Of the 20 well-known buffers proposed by Good, all but three form metal-ion complexes that can interfere with some analyzes.

Under other circumstances, the presence of such an additional charge on the buffer can be a disadvantage. Buffers for electrophoresis and ion-exchange chromatography ideally should have the minimum number of charges, to minimize the ionic strength of solutions (Equation 3.27). Those for ion-exchange chromatography best possess only one species with a charge opposite in sign to that of the exchanger, so that equilibration of the exchanger can be followed by pH changes. Under the appropriate conditions, a buffer cation can absorb to, or desorb from, the ion exchanger only by exchange with H^+, which changes the pH of the effluent solution.

Avoiding interferences from Good's buffers: a contiguous series of noncomplexing tertiary amine buffers covering the entire range of pH 3–11. Q. Yu *et al.* (1997) *Anal. Biochem.* **253**, 50–56.

Oxidation of Good's buffers by hydrogen peroxide. G. Zhao & N. D. Chasteen (2006) *Anal. Biochem.* **349**, 262-267.

4. *Volatile Buffers*

Many separation methods, such as chromatography and electrophoresis, are most convenient if a volatile buffer is used that can be removed afterward by lyophilization and drying. In general, the neutral form of a buffer molecule may be volatile, whereas the charged form will not, because it interacts strongly with water. **A buffer produced by mixing a volatile acid with a volatile base will be volatile only if the two have similar pK_a values.** An example is a mixture of pyridine (pK_a 5.2) and acetic acid (pK_a 4.8). Some of the acetic acid will dissociate and protonate some of the pyridine to the acid form, pyridinium:

$$CH_3{-}CO_2H + C_5H_5N \rightleftharpoons CH_3{-}CO_2^- + C_5H_5NH^+ \qquad (3.20)$$

Acetic acid   Pyridine   Acetate   Pyridinium

The pH will be given by:

$$pH = 5.2 + \log\frac{[\text{pyridine}]}{[\text{pyridinium}]} = 4.8 + \log\frac{[\text{acetate}]}{[\text{acetic acid}]} \qquad (3.21)$$

As the buffer dries, the water, pyridine and acetic acid will evaporate, but they will be regenerated as the reaction in Equation 3.20 is displaced to the left, until all has evaporated.

Ammonium carbonate also forms a useful volatile buffer, as ammonia and carbon dioxide are highly volatile in solution, but it forms a stable crystal lattice when dry, so removing it usually requires several hydration and drying steps. A mixture of triethylamine and carbon dioxide is much more readily volatile. Ammonium acetate and ammonium formate buffers have been used in this way, but the pK_a value of ammonia is so far apart from that of acetic or formic acids that there is very little of the un-ionized forms at equilibrium, so the mixtures are not readily volatile.

Fundamentals of freeze-drying. S. L. Nail *et al.* (2002) *Pharm. Biotechnol.* **14**, 281–360.

3.5. Salts and Ions

Salts interact very strongly with water, and most ions in solution are surrounded by water molecules with varying degrees of order, depending upon their size and charge. Consequently, they behave in aqueous solution as if they are much larger than the ions themselves (Table 3-8). The **hydration number** of an ion refers to the average number of water molecules that are bound sufficiently strongly to it to be removed from the solvent and become part of the solute. Deviations from ideal behavior for concentrated solutions of ions are commonplace, but ideal behavior is demonstrated when mole fractions are calculated by taking account of the tightly bound water. The hydration interaction is primarily electrostatic, between the charge of the ion and the dipole of the water molecule (Figure 3-15). Some of the charge of the ion is believed to be transferred to the solvating water molecules, which increases their hydrogen bonding potential, and the solvated ions should be viewed as hydrogen-bond donors in addition to point charges. The smallest ions tend to have the strongest interactions with water: with its +2 charge and small radius (0.65 Å), Mg^{2+} tightly orders six water molecules in its first or inner hydration shell in an octahedral arrangement. A second, and perhaps even third layer, of water is also organized by the ion charge. To remove these water molecules in isolation would require 455 kcal/mol of free energy. In contrast, K^+ is larger (1.3 Å) and has only a +1 charge. As a result, eight or nine water molecules pack around the ion in a less well-ordered manner, and the hydration energy is only 80 kcal/mol.

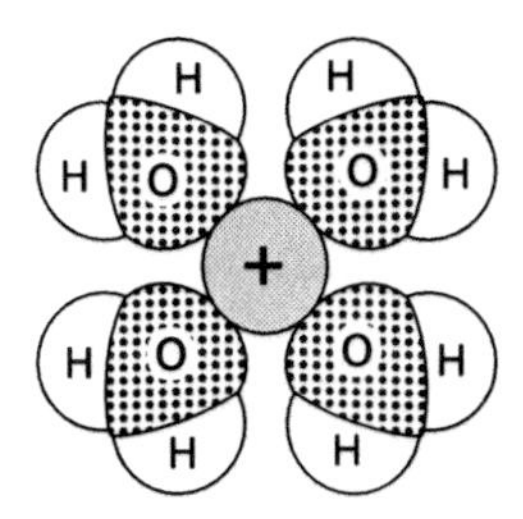

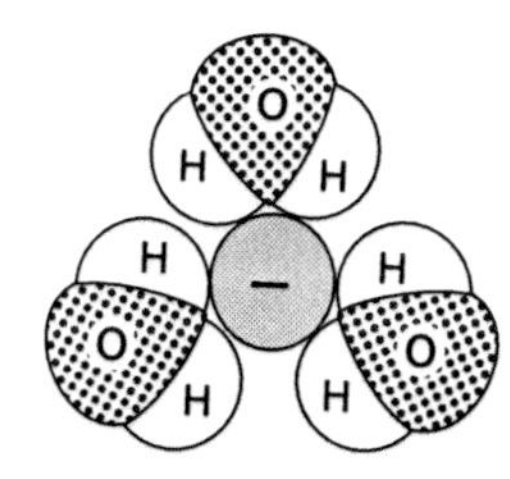

Figure 3-15. Simple schematic diagram of the interaction of cations (+) and anions (–) with surrounding water. Note that this electrostatic interaction should increase the electrostatic dipole of the water molecules and the tendency of the outer atoms of the water molecules to participate in further hydrogen bonds.

Anions tend to interact more strongly with water molecules than do cations, probably because the partial positive charge on the H atoms of the water molecule is on its surface, where it can be closer to the ion, whereas the corresponding negative charge of the O atom is more buried in the center and more distant from an ion (Figure 3-2). This interaction between an ion and water should also increase the dipoles of the water molecules and enhance their participation in hydrogen bonding.

Table 3-8. Comparison of the radii of ions in crystal structures and as inferred from their hydrodynamic properties

	Radius (Å)	
Ion	**Crystal**	**Hydrodynamic**
Li^+	0.6	3.7
Na^+	0.95	3.3
Mg^{2+}	0.65	4.4
Ca^{2+}	0.99	4.2
Zn^{2+}	0.74	4.4

Data from M. Daune (1999) *Molecular Biophysics: structures in motion*, Oxford University Press, Oxford, p. 324.

At least three layers of water and other molecules can usually be distinguished around the central ion, but the first hydration shell is much more important than the others. Some ions have a relatively rigid and stable primary solvation shell, which can be either tetrahedral (Li^+) or octahedral (Mg^{2+}, Co^{2+}, Ni^{2+}) or both (Zn^{2+}). The primary solvation shell of Mg^{2+} is formed by six water molecules with octahedral geometry; the O atoms are 2.07 Å away from the ion. The 12 H atoms of the water molecules are predicted to lie in the equatorial planes of the octahedron. Other monovalent cations and Ca^{2+} have a much more variable first solvation shell, in terms of the number and distance of the water molecules. For example, the Ca^{2+} ion has 12 water molecules bound at 0° C, but only 6.7 on average at 100° C. In general, the number of associated water molecules increases in parallel with the size of the ion, and the ion–water interactions become weaker and more variable.

Anions tend to have the water molecules in the first shell strongly oriented, with one O–H vector pointing directly towards the anion on average, with an angular spread of approximately ±10° for F^-, increasing to approximately ±22° for I^-. The K^+ ion has strong orientational correlations in the first hydration shell, with the water molecules lying with their dipole moments pointing almost directly away from the cation on average, but with an angular spread of approximately ±60°.

The second and third solvation shells are much more variable, and there are also direct hydrogen-bonding interactions between neighboring water molecules.

Aqueous solutions of calcium ions: hydration numbers and the effect of temperature. A. A. Zavitsas (2005) *J. Phys. Chem.* B**109**, 20636–20640.

Ion hydration: implications for cellular function, polyelectrolytes, and protein crystallization. K. D. Collins (2006) *Biophys. Chem.* **119**, 271–281.

Ion solvation and water structure in potassium halide aqueous solutions. A. K. Soper & K. Weckstrom (2006) *Biophys. Chem.* **124**, 180–191.

Coordination numbers of alkali metal ions in aqueous solutions. S. Varma & S. B. Rempe (2006) *Biophys. Chem.* **124**, 192–199.

X-ray and neutron scattering studies of the hydration structure of alkali ions in concentrated aqueous solutions. S. Ansell *et al.* (2006) *Biophys. Chem.* **124**, 171–179.

3.6. ELECTROSTATIC INTERACTIONS IN WATER: DEBYE AND HÜCKEL

All electrostatic interactions in water are diminished by its high dielectric constant, which results from the tendency of the large dipole moments of water molecules to align with any electric field. The dielectric constant of pure water at 0° C is 88.0, so electrostatic interactions in water are only 1.1% the magnitude they would have in a vacuum. Water's dielectric constant decreases monotonically at higher temperatures because thermal motion overcomes the orientational effects of the water dipoles. Its value is 78.54 at 25° C, 75.00 at 35° C and 69.94 at 50° C, so **electrostatic interactions in water become stronger with increasing temperature.**

When small diffusible ions, such as Na^+ and Cl^-, are included in water, the apparent dielectric constant of the solution increases, because **the ions tend to concentrate in the vicinity of charges of the opposite sign**. This **Debye–Hückel screening** can be analyzed by considering the different concentrations of ions at varying distances from a reference ion as being an equilibrium in which the equilibrium constant is determined by the energies of the electrostatic interactions. In this case, n_i, the number of ions of type i per unit volume in a particular region of space, is given by:

$$n_i = n_i^0 \exp(-q_i \varphi / k_B T) \tag{3.22}$$

where n_i^0 is the number density of ions of type i in the bulk solution, q_i is the charge on the ion, φ is the electrostatic potential in that region of space, k_B is Boltzmann's constant and T is the temperature.

This situation is often described by an effective dielectric constant that increases with increasing distance, r, between the charges:

$$D^{eff} = D_{H2O} \exp(+\kappa r) \tag{3.23}$$

where κ is a parameter that is proportional to the square root of the ionic strength (Equation 3.27). The effective dielectric constant increases dramatically with distance, which simply means that electrostatic interactions approach zero. The parameter $1/\kappa$ is known as the **Debye screening distance**, a measure of the distance over which electrostatic effects are damped out by the mobile ions; it is often considered to be the thickness of the ion cloud around each ion. At physiological temperature and ionic strength, about 150 mM, the Debye distance for singly charged ions is about 8 Å, but it varies from about 300 Å in 10^{-4} M salt to only 3 Å at 1 M.

Equation 3.22 only considers the interactions between pairs of ions, so it applies only to dilute solutions where only pairs of ions tend to be in close proximity. To put this in perspective, consider that a 0.1 M solution of a salt like NaCl will have, on average, one ion at the center of a cube with an edge of 20.2 Å, or at the center of a sphere with radius 14 Å. The ions will then tend to be 20–40 Å apart.

Generalizing the Debye–Huckel equation in terms of density functional integral. H. Frusawa & R. Hayakawa (2000) *Phys. Rev. E* **61**, R6079–6082.

Asymmetric primitive-model electrolytes: Debye–Huckel theory, criticality, and energy bounds. D. M. Zuckerman *et al.* (2001) *Phys. Rev. E* **64**, 011206.

Corrected Debye–Huckel analysis of surface complexation. III. Spherical particle charging including ion condensation. M. Gunnarsson *et al.* (2004) *J. Colloid Interface Sci.* **274**, 563–578.

3.6.A. Poisson and Boltzmann

Most calculations of the electrostatic interactions of macromolecules in water use the **Poisson–Boltzmann equation.** This treats the macromolecule as one component, and its interaction with the aqueous salt solution is like that in Equation 3.22. The overall charge density is obtained by adding the contribution from the charges of the macromolecule to the contribution from all the dissolved ions:

$$\nabla \bullet \varepsilon(r)\nabla\varphi(r) = \rho_{\mathrm{macro}}(r) + \sum_i q_i n_i^0 \exp(-q_i\varphi(r)/k_{\mathrm{B}}\mathrm{T}) \tag{3.24}$$

$\rho_{\mathrm{macro}}(r)$ is the charge density due to the macromolecule, the sum of exponentials is the charge density due to the dissolved ions, and ∇ is the differential operator:

$$\nabla = (\partial/\partial \mathrm{x})\,\mathbf{i} + (\partial/\partial \mathrm{y})\,\mathbf{j} + (\partial/\partial \mathrm{z})\,\mathbf{k} \tag{3.25}$$

To use this equation, the structure of the macromolecule and the positions (i,j,k) of all its ionized groups must be known, and the ionic composition of the solvent must be specified. The equation can be solved only by numerical integration. If the electrostatic interactions are relatively weak, where the charged groups are not in close proximity and the ionic strength is relatively low, a linearized approximation of Equation 3.24 is often used:

$$\nabla \bullet \varepsilon(r)\,\nabla\,\phi(r) \;=\; \rho_{macro}(r) + \sum_i q_i\, n_i^0 \tag{3.26}$$

The computations are considerably more straightforward with the linearized version.

In any case, calculations of electrostatic interactions involving the Poisson–Boltzmann equation are generally attempted only by specialists.

Influence of the solvent structure on the electrostatic interactions in proteins. A. Rubinstein & S. Sherman (2004) *Biophys. J.* **87**, 1544–1557.

Incorporation of excluded-volume correlations into Poisson–Boltzmann theory. D. Antypov *et al.* (2005) *Phys. Rev. E* **71**, 061106.

The dependence of electrostatic solvation energy on dielectric constants in Poisson–Boltzmann calculations. H. Tjong & H. X. Zhou (2006) *J. Chem. Phys.* **125**, 206101.

3.7. SOLUBILITIES IN WATER

To be soluble in water, a molecule must occupy the required volume, thereby disrupting the water structure at least within that volume, and possibly nearby as well. The volume that is occupied by such a molecule in solution, its partial molecular volume, reflects not only the van der Waals volume of the molecule but also any changes it causes by rearranging the liquid around it. The solubility of a molecule in water depends upon how many of the unfavorable aspects of creating a cavity in water are compensated for by favorable interactions with the surrounding water molecules (i.e. its hydrophilicity; Section 3.2.A).

In general, **the more polar the surface of a macromolecule, the greater its solubility in water.** On the other hand, the solubility of a molecule depends upon its free energy not only when in solution but also when in whatever solid state it adopts. If its solid state is even more favorable energetically than when in solution, due to strong interactions between the molecules, it will not be very soluble in water. This dependence upon the nature of the solid state makes predicting the solubility of any molecule, especially proteins, very risky.

3.7.A. Salting In, Salting Out

The solubilities of molecules in water generally depend upon the salt concentration. Salt concentrations are frequently expressed as the **ionic strength,** which is defined as:

$$I = \frac{1}{2}\sum_i c_i Z_i^2 \tag{3.27}$$

where c_i is the molar concentration of ion *i* and Z is its ionic charge. The concept of the ionic strength is intended to normalize the electrostatic effects of ions with different charges, but it does not account for the varying other effects of the different ions (Section 3.8).

In general, the dependence of the solubility of a molecule on salt concentration is one in which, for increasing concentrations of any salt, **the solubility increases at low salt concentrations, then passes through a point of maximum solubility, after which further addition of salt reduces the solubility, to an extent depending upon the nature of the salt** (Figure 3-16). This complex behavior has been studied most extensively with water-soluble proteins and is attributed to two different phenomena. At low salt concentrations, proteins are **salted-in,** i.e. their solubility increases. For any protein, this is independent of the salt and is simply a consequence of the protein carrying positively and negatively charged groups on its accessible surface. These groups are surrounded by an atmosphere of the ions of the salt, which increases the solubility according to the simple equation:

$$\log(S_s/S_w) = Z^2\, I^{1/2}\, A \tag{3.28}$$

where S_s and S_w are the solubility in the presence and absence of salt, respectively, Z is the net charge of the protein, I is the ionic strength and A is a combination of constants, some of which are specific for the protein.

At high salt concentrations, greater than approximately 0.5 M, the effect of **salting-out** predominates. The protein solubility then decreases with increasing salt concentration according to the equation:

$$\log S_s = \beta - K^s\, I \tag{3.29}$$

where β is an empirical constant that should be specific for each protein and K^s is the **salting-out constant,** which is characteristic of the particular salt (Figure 3-17). The order of salting-out effectiveness of the various salts is very similar for most proteins and follows the Hofmeister series (Section 3.8). The magnitude of K^s is determined by the preferential binding of the particular salt with the protein (Section 3.9.A). Salts that induce preferential hydration, i.e. negative binding of the salt, reduce the solubilities of proteins, whereas a salt that exhibits preferential binding increases the solubility.

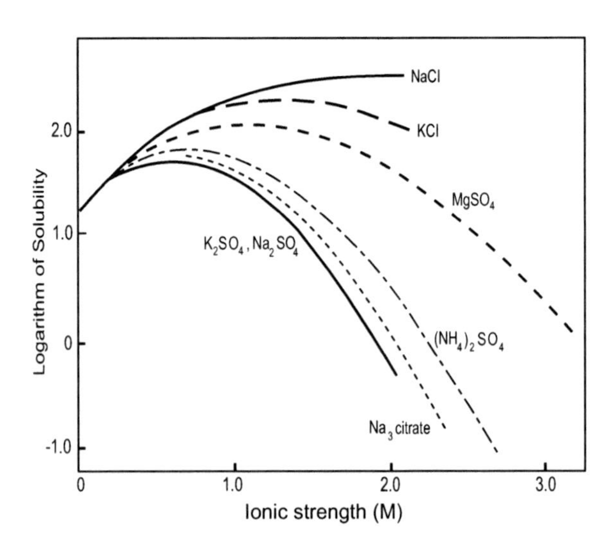

Figure 3-16. The solubility of hemoglobin (with carbon monoxide bound) in various electrolytes at 25° C. The solubility is expressed as grams per 1000 grams of H_2O. Data from A. A. Green.

Apparent Debye–Huckel electrostatic effects in the folding of a simple, single domain protein. M. A. de Los Rios & K. W. Plaxco (2005) *Biochemistry* **44**, 1243–1250.

A molecular-thermodynamic model for the interactions between globular proteins in aqueous solutions. L. Jin *et al.* (2006) *J. Colloid Interface Sci.* **304**, 77–83.

Aqueous salting-out effect of inorganic cations and anions on non-electrolytes. M. Gorgenyi *et al.* (2006) *Chemosphere* **65**, 802–810.

A fluctuation theory analysis of the salting-out effect. R. M. Mazo (2006) *J. Phys. Chem.* **110**, 24077–24082.

Effect of salts and organic additives on the solubility of proteins in aqueous solutions. E. Ruckenstein & I. L. Shulgin (2006) *Adv. Colloid Interface Sci.* **123**, 97–103.

Quantification and rationalization of the higher affinity of sodium over potassium to protein surfaces. L. Vrbka *et al.* (2006) *Proc. Natl. Acad. Sci. USA* **103**, 15440–15444.

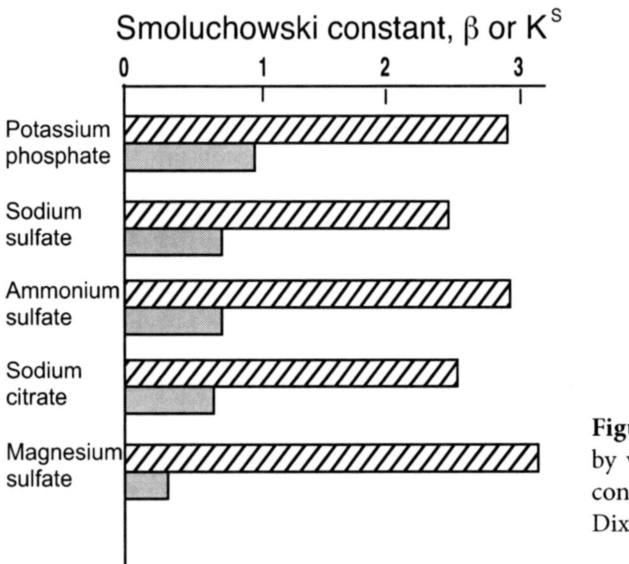

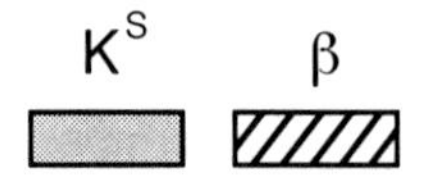

Figure 3-17. The salting-out of carboxyhemoglobin by various salts as measured by the Smoluchowski constants β and K^S of Equation 3.29. Data from M. Dixon & E. C. Webb.

3.8. HOFMEISTER SERIES

The Hofmeister series, or **lyotropic series**, was first described by Hofmeister in 1888 as part of his work on the effectiveness of salts in precipitating serum globulins. This same ordering of ions in their effectiveness in salting-out proteins has been encountered over and over again in a variety of phenomena, including the effects of these ions on the stabilities of both nucleic acids and proteins, the association/dissociation equilibria of macromolecules, enzyme activity and various other biochemical functions, plus the solubilities of small molecules and even measurement of the pH of solutions. The ordering is as follows:

$$\leftarrow \text{Stabilization} \qquad \text{Destabilization} \rightarrow \tag{3.30}$$

$$\leftarrow \text{Salting out} \qquad \text{Salting in} \rightarrow$$

$$\text{Anions: } SO_4^{2-} > CO_3^{2-} > F^- > CH_3COO^- > Cl^- > Br^- > NO_3^- > ClO_4^- > I^- > SCN^-$$

$$\text{Cations: } NH_4^+ > Rb^+ = K^+ > Na^+ = Cs^+ > Li^+ > Mg^{2+} > Ca^{2+} > Ba^{2+} > Gdm^+$$

Anions and cations are essentially independent of each other in these actions, and their effects are additive. For example, LiBr is a salting-in agent, while KF is a salting-out agent. Reshuffling the ions gives the result that both KBr and LiF have no effect on the solubility; i.e., the salting-in capacity of one ion (Li^+ or Br^-) is compensated by the salting-out characteristic of the co-ion (F^- or K^+). Similarly, guanidinium ion (Gdm^+) is a strong salting-in and destabilizing ion, and GdmCl (also known as guanidine hydrochloride) is a strong denaturant. But GdmSCN is a substantially stronger denaturant, whereas Gdm_2SO_4 actually stabilizes proteins; the stabilizing effect of sulfate outweighs the destabilizing effect of the guanidinium ion. The anion tends to be more influential than the cation.

The ions on the left of each series tend to be small or to have high charge densities. They are believed to augment the structure of the water and are often described as **kosmotropes**. They decrease the solubilities of nonpolar molecules (i.e. salt out) and increase the surface tension of water, which may increase the energy required to form a cavity in water; in effect, they strengthen the hydrophobic interaction. In contrast, the ions on the right of each series tend to have low charge densities. They are believed to disrupt the structure of water and are known as **chaotropes**. They increase the solubilities of nonpolar molecules (i.e. salt in) and in effect weaken the hydrophobic effect. The dividing points between the two effects are usually taken as Na^+ and Cl^-; NaCl is approximately neutral in this respect.

The mechanism of action of these ions is believed to be related to their hydration, i.e. their effects on the orientation of water molecules, through electrostatic effects on their hydrogen-bond donor and acceptor properties. A major part of their effect arises because they are excluded to varying extents from interfaces, such as that of water with air or with a nonpolar surface. Exclusion of the co-solvent from the air–water interface is the reason for the increase in surface tension of the solution. The water structure at interfaces with gas or nonpolar surfaces is already perturbed by the nonavailability of hydrogen-bonding groups for water molecules at the surface (Section 3.2). This surface is crucial for determining the aqueous solubility. The correlation between the potency of Hofmeister salts and their effects on surface tension arises from partitioning of anions and cations between bulk water and the air–water interface. Most anions that favor processes that expose protein surface to water (e.g.

SCN^-), and hence must interact favorably with protein surfaces, also accumulate preferentially at the air–water interface. Most anions that favor processes that remove protein surface from water (e.g. F^-), and hence are excluded from a protein surface, are also excluded from the air–water interface. The guanidinium cation, a strong protein denaturant and therefore preferentially bound to the protein surface exposed in unfolding, is somewhat excluded from the air–water surface but is much less excluded than alkali metal cations (e.g. Na^+ and K^+). Hence cations appear to interact more favorably with protein surfaces than with the air–water interface, which might be due to the hydrogen-bonding interactions between some of them and the protein.

Interactions of macromolecules with salt ions: an electrostatic theory for the Hofmeister effect. H. X. Zhou (2005) *Proteins* **61**, 69–78.

Is there an anionic Hofmeister effect on water dynamics? Dielectric spectroscopy of aqueous solutions of NaBr, NaI, $NaNO_3$, $NaClO_4$, and NaSCN. W. Wachter *et al.* (2005) *J. Phys. Chem. A* **109**, 8675–8683.

Why pH titration in protein solutions follows a Hofmeister series. M. Bostrom *et al.* (2006) *J. Phys. Chem. B* **110**, 7563–7566.

The Hofmeister series and protein–salt interactions. S. Shimizu *et al.* (2006) *J. Chem. Phys.* **124**, 234905.

Interactions between macromolecules and ions: the Hofmeister series. Y. Zhang & P. S. Cremer (2006) *Curr. Opinion Chem. Biol.* **10**, 658–663.

Hofmeister salt effects on surface tension arise from partitioning of anions and cations between bulk water and the air–water interface. L. M. Pegram & M. T. Record (2007) *J. Phys. Chem. B* **111**, 5411–5417.

3.9. HYDRATION OF MACROMOLECULES

The functions and properties of nearly all biological macromolecules are governed by how they interact with water. Either they are soluble in it, or they prefer to interact with other macromolecules or with membranes. The interactions of a macromolecule with solvent are determined primarily by its surface. **The total hydration of a macromolecule is the effective amount of water immobilized by, or bound to, the macromolecule.** It is the sum of the interactions of water molecules with individual sites on the accessible surface of the molecule. Such interactions vary from:

(i) very strong interactions, such as water molecules trapped within cavities in a protein molecule and involved in the actual folded structure;

(ii) weak interactions, such as water molecules hydrating charged and other polar groups on the surface;

(iii) to very weak interactions that comprise water molecules whose rotation or translation is momentarily perturbed by their proximity to the macromolecule.

This last type of interaction is primarily entropic in nature and has a relatively small free energy. All these interactions cause an effective binding of water, i.e. immobilization of the water by the macromolecule. Such binding may be present only a small fraction of the time at each site, but it can add up to a substantial degree of binding over the entire surface at any instant of time.

Total hydration is very difficult to measure, and its value may be a function of the techniques used. It is also very difficult to understand the details of the interactions between macromolecule and solvent, because they are largely invisible. X-ray crystallography detects only the more strongly interacting water molecules that are held in one position; the others occur in so many positions that their electron density averaged over time and the many molecules of the crystal is smeared out into a continuum. NMR can detect water molecules whose freezing is perturbed by the presence of the macromolecule, but the great majority of hydrating waters are not perturbed noticeably in their NMR properties.

Hydration theory for molecular biophysics. M. E. Paulaitis & L. R. Pratt (2002) *Adv. Protein Chem.* **62**, 283–310.

Determination of the enthalpy of solute–solvent interaction from the enthalpy of solution: aqueous solutions of erythritol and L-threitol. A. J. Lopes *et al.* (2006) *J. Phys. Chem. B* **110**, 9280–9285.

Pressure and temperature dependence of hydrophobic hydration: volumetric, compressibility, and thermodynamic signatures. M. S. Maghaddam & H. S. Chan (2007) *J. Chem. Phys.* **126**, 114507.

Solvation free energy of amino acids and side-chain analogues. J. Chang *et al.* (2007) *J. Phys. Chem. B* **111**, 2098–2106.

3.9.A. Preferential Hydration versus Preferential Binding

A major question is how the co-solutes of an aqueous solution interact with the surfaces of biological macromolecules and thereby affect their solubilities and stabilities. They can interact either less or more favorably than water of the bulk solvent, and they are usually treated as a ligand of the macromolecule.

The equilibrium binding of a ligand (L) to a macromolecule (M) at a specific site can be expressed by a simple mass-action equation:

$$\mathrm{L} + \mathrm{M} \leftrightarrow \mathrm{L} \bullet \mathrm{M} \tag{3.31}$$

with the equilibrium constant for binding, K_b:

$$K_b = \frac{[\mathrm{L} \bullet \mathrm{M}]}{[\mathrm{L}][\mathrm{M}]} \tag{3.32}$$

The binding isotherm describes the average number of moles of ligand bound per mole of macromolecule, $\bar{\nu}$, as a function of the concentration of free ligand, and is given by:

$$\bar{\nu} = \frac{nK_b[\mathrm{L}]}{1 + K_b[\mathrm{L}]} \tag{3.33}$$

where n is the number of sites.

Extremely weakly interacting ligands, however, such as the agents that affect protein stability and solubility, are required at high concentrations, > 1 M. Consequently, they are more properly considered as co-solvents, **and displacement of water from the binding sites must be taken into account explicitly** (Figure 3-18). The water–ligand exchange at a site is described by:

$$M \bullet qH_2O + L \leftrightarrow L \bullet M + qH_2O \tag{3.34}$$

where q is the average number of water molecules displaced by L; it need not be an integer.

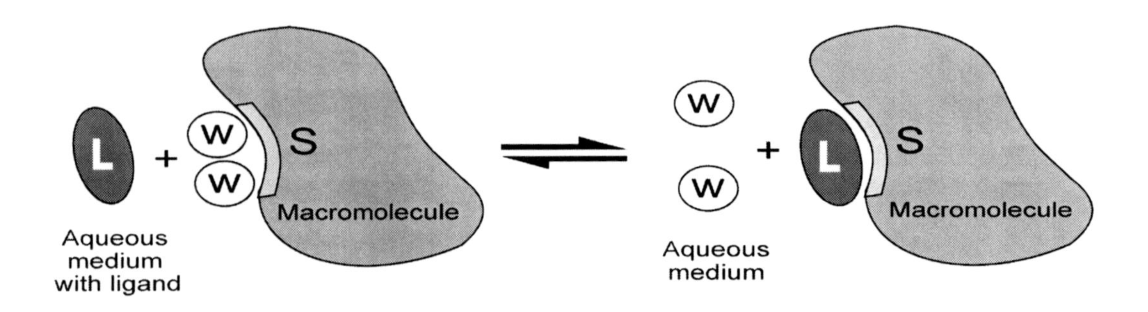

Figure 3-18. Schematic representation of the replacement of water molecules (W) by a ligand (L) at a binding site (S) on a macromolecule. Adapted from S. N. Timasheff.

The measured binding equilibrium constant becomes an exchange constant:

$$K_b = K_{ex} = \frac{[L \bullet M][H_2O]^q}{[L][M \bullet qH_2O]} \tag{3.35}$$

In the most simple case, where one ligand molecule displaces one water molecule, the binding isotherm is:

$$\bar{\nu} = \frac{n(K_{ex} - \frac{1}{m_W})m_L}{1 + K_{ex}\, m_L} \tag{3.36}$$

where m_W and m_L are the molal (moles per 1000 g H_2O) concentrations of water and ligand, respectively. Since m_W = 55.56 moles H_2O per 1000 g H_2O, $(1/m_W)$ = 0.018 m^{-1}. Consequently, the experimental results of a binding measurement (e.g. by dialysis equilibrium), can give negative as well as positive values of the extent of binding. The measured binding is positive when $K_{ex} > 0.018\ m^{-1}$ and negative when $K_{ex} < 0.018\ m^{-1}$. The extent of preferential binding, of any binding, weak or strong, is usually obtained from dialysis equilibrium measurements of the ligand concentration inside and outside the dialysis bag:

$$\bar{\nu} = \frac{[\text{Ligand inside bag}] - [\text{Ligand outside bag}]}{[\text{Macromolecule}]} \tag{3.37}$$

In other words, the magnitude and sign of $\bar{\nu}$ arise simply from the difference between two measured concentrations. **Many substances that stabilize protein structure give negative values of binding in this way**. For example, dialysis equilibrium measurements of the protein ribonuclease A (RNaseA) in 1 M sucrose at pH 7 indicate that −7.6 moles of sugar are bound per mole of the protein, i.e. there is a deficiency of sugar molecules on the protein surface relative to the sugar concentration in the bulk solvent (Figure 3-19). This is referred to as **preferential exclusion**. An insufficiency of sugar at the protein surface implies that there must be an excess of water, which is **preferential hydration**.

The number of molecules of a solvent component, water or co-solvent, that form contacts with the macromolecular surface is **total binding**. Total binding can be measured by techniques that respond to contacts between ligand molecules and the macromolecule, such as calorimetric titration (Figure 1-4), which detects the heat of protein–ligand contact, and spectroscopic techniques, such as fluorescence and UV absorbance, which detect spectral perturbations each time a contact between ligand and macromolecule occurs.

Binding measured by most other techniques is **preferential binding**. It is related to total binding, B_L, by:

$$\bar{\nu} = B_L - \frac{m_L}{m_W} B_W \qquad (3.38)$$

where B_W is the binding of water to the surface. B_L and B_W are the numbers of molecules of ligand and water that are in contact with the surface of the protein molecule at any moment. Most binding experiments give the true extent of binding of a ligand, B_L, only if the second term of Equation 3.38 is negligible. This is the case with low concentrations of ligands, which bind tightly at a few specific sites on a macromolecule, but not with co-solvents that must be used at high concentrations.

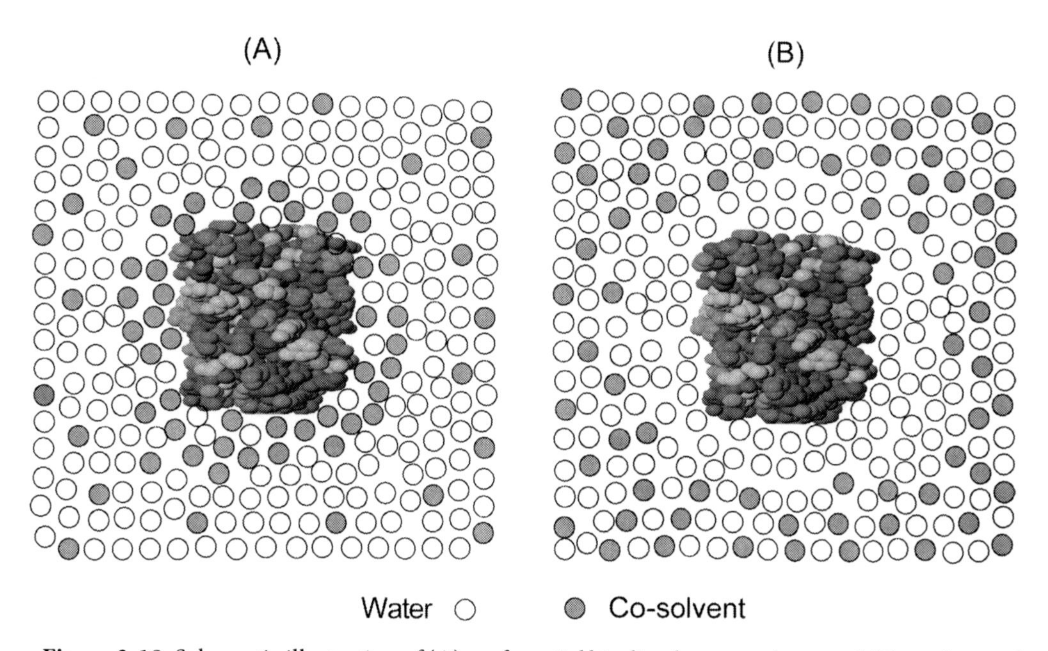

Figure 3-19. Schematic illustration of (A) preferential binding by a co-solvent and (B) preferential hydration. Preferential binding causes the co-solvent to be present in the solvation shell of the macromolecule at higher concentrations than in the bulk solvent. In preferential hydration, the co-solvent is excluded from the hydration shell and there is an enrichment of water. Adapted from G. C. Na & S. N Timasheff.

Preferential hydration and the exclusion of co-solvents from protein surfaces. S. Shimizu & D. J. Smith (2004) *J. Chem. Phys.* **121**, 1148–1154.

Preferential hydration of DNA: the magnitude and distance dependence of alcohol and polyol interactions. C. Stanley & D. C. Rau (2006) **91**, 912–920.

Molecular level probing of preferential hydration and its modulation by osmolytes through the use of pyranine complexed to hemoglobin. C. J. Roche *et al.* (2006) *J. Biol. Chem.* **281**, 38757–38768.

On the nature of ions at the liquid water surface. P. B. Petersen & R. J. Saykally (2006) *Ann. Rev. Phys. Chem.* **57**, 333–364.

A protein molecule in a mixed solvent: the preferential binding parameter via the Kirkwood–Buff theory. I. L. Shulgin & E. Ruckenheim (2006) *Biophys. J.* **90**, 704–707.

3.9.B. Transfer Free Energy

Interactions of co-solvents with macromolecules can also be measured quantitatively from the free energy of transferring the macromolecule from water to a solution of the co-solvent. The transfer free energy of a protein, $\Delta\mu_{Pr.tr}$, is the change in free energy of transferring the protein from pure water into an aqueous solvent containing a co-solvent, such as 1 M sucrose, 8 M urea or 6 M GdmCl. It is a measure of the change in the free energy of interaction of the protein with the new solvent, i.e. the free energy of binding of the co-solvent to the protein, ΔG_b.

When the preferential binding measured by dialysis equilibrium (Equation 3.37) is expressed by the perturbation of the chemical potential of the protein, μ_{pr}, by the co-solvent, or ligand, L:

$$(\partial\mu_{Pr}/\partial m_L)_{T,Pr} = -(\partial m_L/\partial m_{Pr})_{T,Pr,L}\,(\partial\mu_L/\partial m_L)_{T,Pr} \tag{3.39}$$

where μ_L is the chemical potential of the ligand or co-solvent, m_{Pr} and m_L are the molal concentrations of the protein and ligand, respectively, and the subscripts T, Pr and L outside the parentheses indicate that the temperature and the protein and ligand concentrations are kept constant. The term $(\mu_L/\partial m_L)_{T,Pr}$ is the nonideality of the ligand or co-solvent and is given by:

$$(\partial\mu_L/\partial m_L)_{T,Pr} = RT\left(\frac{1}{m_L} + \partial\log_e\gamma_L/\partial m_L\right) \tag{3.40}$$

where γ_L is the activity coefficient of the co-solvent, R is the gas constant and T the temperature.

The transfer free energy can be obtained by integration over the co-solvent concentration:

$$\Delta\mu_{Pr} = \int_0^L (\partial\mu_{Pr}/\partial\mu_L)_{T,Pr}\,dm_L = \Delta G_b \tag{3.41}$$

3.10. CHEMICAL POTENTIAL

In multiple-component systems, it is necessary to consider the effects of changes in concentration of each of the components: in thermodynamic terms, to consider the Gibbs free energy of a system,

G (Section 1.2), in terms of the components that are present. This is usually done by defining the **chemical potential** of each component i, μ_i, as the change in free energy of the entire system as the concentration of component i, n_i, increases:

$$\mu_i = \partial G/\partial n_i \quad (3.42)$$

The chemical potential is essentially the partial free energy of the component.

If both the temperature and pressure are kept constant, the free energy of the system depends only on its composition and the chemical potentials of its various components:

$$G = \Sigma_i\, n_i \mu_i \quad (3.43)$$

At constant temperature and pressure, if the chemical potential of one component changes, that of the others must also change coincidentally in some fashion, because the sum of the changes in chemical potentials must remain zero:

$$\Sigma_i\, n_i \mathrm{d}\mu_i = 0 \quad (3.44)$$

Chemical potential derivatives and preferential interaction parameters in biological systems from Kirkwood–Buff theory. P. E. Smith (2006) *Biophys. J.* **91**, 849–856.

The effect of salt on protein chemical potential determined by ternary diffusion in aqueous solutions. O. Annunziata *et al.* (2006) *J. Phys. Chem.* **110**, 1405–1415.

3.11. COMPRESSIBLILTY: THE EFFECTS OF HIGH PRESSURES

Pressure is a fundamental physical parameter, because changing the temperature of a biochemical system at constant pressure causes simultaneous changes in both the thermal energy and the volume. To separate out these effects requires high pressures to keep the volume constant. Increased pressure usually decreases the volume of a system, and its compressibility is an important measure of the atomic packing and flexibility of macromolecules in solution. The compressibility, K, of a system is defined as the negative pressure derivative of its volume, V:

$$K = \beta V = -(\partial V/\partial P) \quad (3.45)$$

When measured at constant temperature, the parameters are designated as 'isothermal' and given the subscript T. In contrast, 'adiabatic' measurements are made at constant entropy and given the subscript S. The two coefficients are related to each other by:

$$\beta_T = \beta_S + (T\alpha^2/\rho C_p) \quad (3.46)$$

where ρ is the density, α the coefficient of thermal expansion and C_p the specific heat capacity at constant pressure (Section 1.5). The **partial specific compressibility** of a macromolecule in solution is defined as the change in the partial specific volume, ν_2^0, of the macromolecule with increasing pressure.

The compressibilities of liquids are most readily measured using the **Newton–Laplace equation**:

$$\beta_S = \rho^{-1} U^{-2} \quad (3.47)$$

where ρ is the density of the medium and U the velocity of sound there. The sound velocity can usually be measured with an accuracy of 1 cm s^{-1} by using a resonance or 'sing-around' pulse method at 3–6 MHz. Most studies of compressibility have been of proteins, because of the many questions posed by their complex structures, and the remainder of this discussion will focus on them.

The van der Waals volume of the atoms of a molecule is undoubtedly not compressible to any significant extent, so the observed compressibility K of a molecule in solution can be divided into two terms: (1) the intrinsic compressibility, K_M ($= \beta_M V_M$), of the cavities within the interior of the molecule, and (2) the compressibility K_h of the hydration shell around the protein. The value of V_M reflects the spatial architecture of the macromolecular interior, and the coefficient β_M the tightness of the internal packing. The isothermal compressibility, β_T, of a protein with volume V_p defines its average fluctuations in volume by:

$$\delta V_p^{\,2} = k_B T V_p \beta_T \quad (3.48)$$

where k_B is Boltzmann's constant and T the absolute temperature.

The hydration terms depend upon the number and nature (charged, polar or nonpolar) of solvent-exposed protein groups. The hydration compressiblities have been measured with small molecules, where the intrinsic compressibility K_M is negligible. All amino acids in solution show large negative β_s, with values between -62.5×10^{-12} cm^2 dyn^{-1} (for glycine) and -21×10^{-12} cm^2 dyn^{-1} (for tryptophan) at 25° C. These measurements have shown that water solvating charged groups occupies a smaller volume due to hydrogen bonding and electrostriction and is less compressible than bulk water at all temperatures, whereas water solvating aliphatic groups is less compressible than bulk water at temperatures below 35° C, but greater at higher temperatures. Water hydrating an isolated polar group is much more compressible than bulk water, but the difference decreases markedly at high temperatures. Water hydrating multiple polar groups in close proximity is less compressible than bulk water and depends only slightly on temperature. The changes in the properties of water induced by the solvation of another molecule occur mainly to the primary layer of water molecules but also extend at least partially to the second layer, and even to the third layer in the case of macromolecules.

Determination of the volumetric properties of proteins and other solutes using pressure perturbation calorimetry. L. N. Lin *et al.* (2002) *Anal. Biochem.* **302**, 144–160.

Isentropic and isothermal compressibilities of the backbone glycyl group of proteins in aqueous solution. G. R. Hedwig (2006) *Biophys. Chem.* **124**, 35–42.

What lies in the future of high-pressure bioscience? C. Balny (2006) *Biochim. Biophys. Acta* **1764**, 632–639.

~ CHAPTER 4 ~

KINETICS: A BRIEF REVIEW

A biological system at equilibrium is dead. Life requires a dynamic system in which change is occurring constantly. Chemical reactions in biological systems are catalyzed by enzymes at microsecond to millisecond time scales, and biochemicals are continually being synthesized, converted to new forms and degraded. The rates at which such changes occur are described by kinetics.

The distinction between thermodynamics (Chapter 1) and kinetics is vital. Thermodynamics is concerned with how much of a substance, S, can be converted into a product, P, at equilibrium, irrespective of the rate, which might be immeasurably slow. In contrast, kinetics is concerned with the rate and the pathway by which the conversion takes place.

Kinetics and Mechanisms. A. A. Frost & R. G. Pearson (1962) John Wiley, NY.

Kinetics and Mechanisms. W. J. Moore & R. G. Pearson (1981) John Wiley, NY.

Kinetics for the Life Sciences. H. Gutfreund (1995) Cambridge University Press, Cambridge.

4.1. SINGLE REACTIONS

The kinetics of reactions are usually followed by introducing at time zero (t = 0) a sample of the reactant to conditions where the reaction can take place. Normally this involves mixing two solutions, one containing the reactant and the other supplying any other reactant required or changing the conditions, such as the pH or temperature, so that the reaction will commence. The time–course of the reaction is then followed by monitoring as a function of time the decrease in the concentration of the original reactant or the increase in concentration of a product of the reaction. Ensembles of many molecules are followed, so the average property of a large population of molecules is observed. For example, 1 ml of a 1 μM solution contains 1 nmol of reactant, or 6×10^{14} molecules.

The velocity of a reaction is defined by the rate of disappearance of a reactant or by the rate of appearance of a product. For the most simple reaction S → P:

$$\text{velocity} = \frac{-\,\mathrm{d}[\mathrm{S}]}{\mathrm{d}t} = \frac{\mathrm{d}[\mathrm{P}]}{\mathrm{d}t} \tag{4.1}$$

The velocity will be measured in units of concentration per unit of time, for example M s^{-1}. The observed rate of any reaction will depend upon the concentration of each of the reactants, such as [S], raised to some power, *i*, which depends upon the number of molecules of S involved in the reaction prior to the rate-determining step. For example, the simple reaction S → P involves only a single molecule of the reactant S, so the rate equation is given by:

$$\text{velocity} = \frac{-\text{d[S]}}{\text{d}t} = k_{app}\,[\text{S}] \tag{4.2}$$

If the reaction involves the simultaneous encounter of *i* molecules of S, the rate equation is given by:

$$\text{velocity} = \frac{-\text{d[S]}}{\text{d}t} = k_{app}\,[\text{S}]^i \tag{4.3}$$

The proportionality constant k_{app} is known as the **apparent rate constant**. Its units will depend upon the order of the reaction: $\text{M}^{(1-i)}\,\text{s}^{-1}$.

Chemical reactions can be described in terms of their **molecularity** and **order**. The molecularity is the number of molecules participating in the overall reaction. Thus the conversion of S → P is a unimolecular reaction, while the reactions S + A → P and S + A + B → P are bimolecular and trimolecular reactions, respectively. The order of the reaction is defined by the power of the concentration-dependence of the rate of conversion of the starting substance S into the products, P, the parameter *i* in Equation 4.3. Reactions are said to be zero-, first-, or second-order with regard to a reactant when *i* is, respectively, 0, 1 or 2; even greater values of *i* are possible theoretically but rarely encountered in practice in molecular biology, because **simultaneous collisions involving more than two molecules are very unlikely in dilute solutions with low concentrations of the reactants.** For a reaction with multiple reactants, the rate of the overall reaction will depend upon some order of each of them:

$$\text{velocity} = \frac{-\text{d[S]}}{\text{d}t} = k_r\,[\text{S}]^i\,[\text{A}]^j\,[\text{B}]^k \tag{4.4}$$

The units of the rate constant will be $\text{M}^{(1-i-j-k)}\,\text{s}^{-1}$. The values of *i*, *j* and *k* are usually integers when the molecules react as individual molecules in a chemical reaction. On the other hand, noninteger values are possible when the reactant merely associates differently with the products and the initial reactants, as can happen with water (Section 3.9). The order of the reaction depends upon which reactants are involved in the rate-determining step of a reaction (Section 4.2.A), which need not be the same as the molecularity of the reaction. A reactant that enters the reaction only after the rate-determining step will not appear in the kinetic equation (Equation 4.4) for the forward reaction, so long as its concentration is adequate. It will, however, appear in the rate equation for the reverse reaction, as its presence will reverse steps before the rate-limiting step in that direction (Equation 4.52).

Reaction kinetics in intracellular environments with macromolecular crowding: simulations and rate laws. S. Schnell & T. E. Turner (2004) *Prog. Biophys. Mol. Biol.* **85**, 235–260.

4.1.A. First-Order Kinetics

First-order kinetics is the most simple, as a single molecule of reactant is involved and its rate of disappearance is dependent on the first power of its concentration:

$$\text{velocity} = \frac{-\,d[S]}{dt} = k_{uni}\,[S] \tag{4.5}$$

Examples of such unimolecular reactions are the isomerization of a molecule (S → P) or the conversion of one radioisotope into another (Chapter 5); in neither case is another molecule necessarily involved. The apparent rate constant, k_{uni}, has units of (time^{-1}) and the velocity has the required units of (concentration per time). Higher order reactions involving other reactants can also exhibit apparently first-order kinetics (**pseudo-first-order**) if the concentrations of the other reactants are present in excess, so that their concentrations remain essentially constant throughout the time–course of the reaction (Equation 4.25).

The progress of a first-order reaction can be predicted by integrating Equation 4.5. If the reaction is irreversible and the starting concentration of S is $[S]_0$, the rate equations are:

$$\frac{-d[S]}{[S]} = k_{uni}\,dt \tag{4.6}$$

$$\log_e \frac{[S]}{[S]_0} = 2.303\,\log_{10}\left(\frac{[S]}{[S]_0}\right) = -k_{uni}\,t \tag{4.7}$$

The velocity of the reaction will decrease with time, as the concentration of S decreases (Figure 4-1). Plotting the natural logarithm of [S] versus *time* according to Equation 4.7 should yield a straight line, with a negative slope equal to the apparent rate constant k_{uni} (Figure 4-2-A).

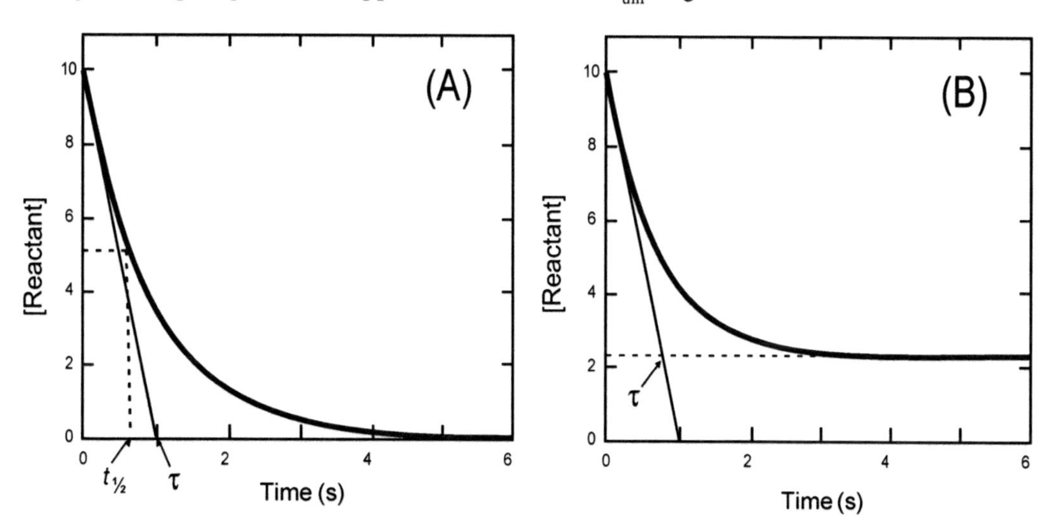

Figure 4-1. Time–course of an irreversible one-step reaction (A) and a reversible one (B). The concentration of the reactant was simulated as a function of time with $k_f = 1$ s^{-1} and $k_r = 0$ (in A) or 0.25 s^{-1} (in B). For the irreversible reaction, the half-time ($t_{1/2}$) is 0.69 s and the relaxation time (τ) is 1.0 s. For the reversible reaction, both the half-time and the relaxation time are decreased: τ is 0.8 s [= $(k_f + k_r)^{-1}$]. The initial rate is the same with the irreversible and reversible reactions, but the reversible reaction proceeds less far, so the apparent half-time and relaxation times are shorter.

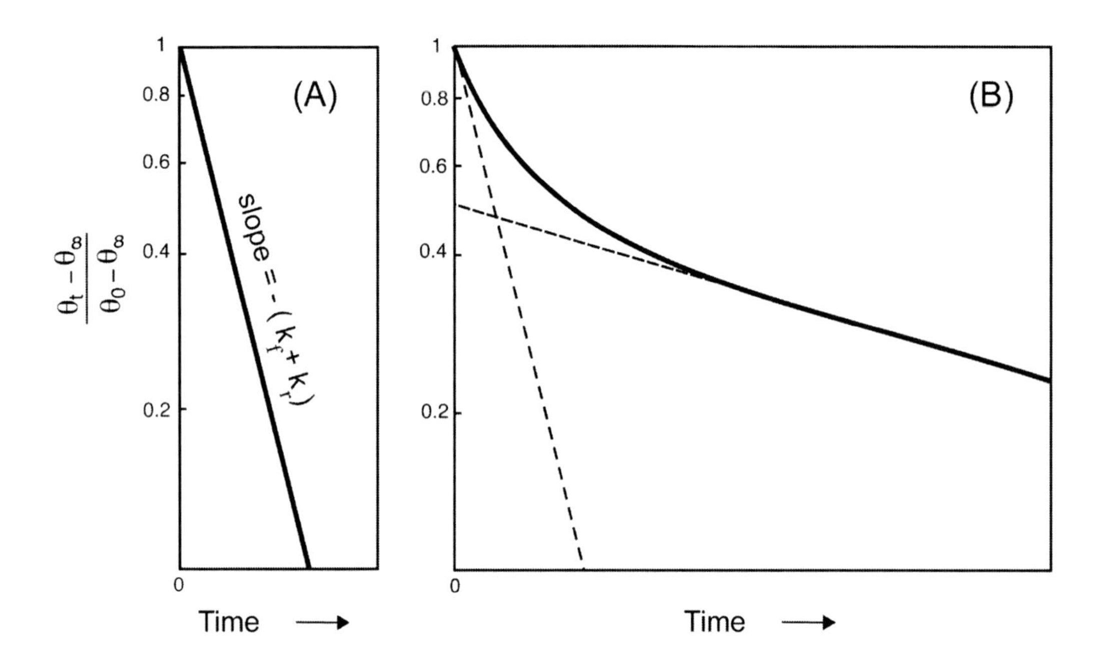

Figure 4-2. Time-course of a first-order reaction when the logarithm of the extent of reaction (θ_t) is plotted as a function of time. The initial and final extents of the reaction are given by θ_0 and θ_∞, respectively. Note that this axis is logarithmic. (A) A single-step reaction; the negative slope of the straight line gives the sum of the forward and reverse rate constants. (B) The biphasic kinetics observed with two phases (indicated individually by the *dashed lines*). Biphasic kinetics result if two different reactions are being followed, with two different rate constants, or if an initial reversible reaction is followed by a slower one, so that an intermediate accumulates. It is not possible to distinguish between these two types of mechanism with solely kinetic data like these.

Equations 4.5–4.7 imply that every molecule of S has the same probability of undergoing the reaction, and the rate constant can be interpreted as such a probability. The time over which the reaction is observed to take place in a large ensemble of molecules does not mean that the reaction of each molecule requires that much time to be completed. Instead, once initiated in a molecule, the reaction is usually completed very quickly, perhaps within 10^{-12} s (Section 4.1.D). The time required for the population of molecules to react depends solely upon the probability of each molecule initiating the reaction. This has been confirmed by watching individual molecules undergo a reaction, which is possible using very sensitive techniques and very dilute solutions of the reactant. In this case, the molecules are observed to do nothing noticeable until they suddenly initiate and complete the reaction. A histogram of the periods of time the individual molecules spend before the rapid reaction occurs mimics the first-order kinetics observed with a large population of molecules (Figure 4-3).

The linear semi-log plot of first-order reactions (Figure 4-2-A) is a stringent requirement. It is relatively easy (so long as the final extent of the reaction is known accurately) to detect departures from such kinetics, which would indicate more complex situations than those assumed above. For example, a reactant might be heterogeneous in some way that will cause the different populations to react at different rates:

$$\begin{aligned} S_1 &\xrightarrow{k_1} P_1 \\ S_2 &\xrightarrow{k_2} P_2 \end{aligned} \qquad (4.8)$$

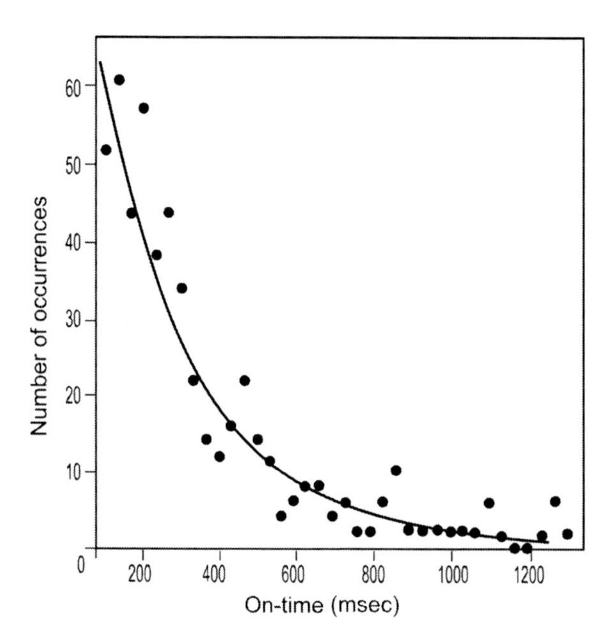

Figure 4-3. The frequency of reaction in a single molecule produces first-order kinetics. A histogram of the times between reactions (*circles*) mimics the exponential decay expected with a first-order reaction involving large numbers of molecules (*solid line*). The reaction was occurring on a single molecule of the enzyme cholesterol oxidase. Data from Lu *et al.* (1998) *Science* **282**, 1877–1881.

In that case, biphasic kinetics will be observed (Figure 4-2-B); each phase will correspond to one of the reactions. Biphasic kinetics of the disappearance of S will also result if the initial reaction is reversible and the product of that reaction is more slowly converted to another product:

$$S \xleftrightarrow{\text{fast}} I \xrightarrow{\text{slow}} P \tag{4.9}$$

In practice, however, two such reactions have to differ in rates by at least a factor of 3 for the two phases to be clearly distinguishable. Such situations will be deciphered most readily if the rates of appearance of the various products can be monitored.

Another possibility is that reactant S might undergo more than one reaction simultaneously:

$$S \xrightarrow{k_P} P \qquad S \xrightarrow{k_Q} Q \tag{4.10}$$

First-order kinetics will be observed, and the apparent rate constant for the disappearance of S will be the sum of all the rate constants of the individual reactions. Such a situation will be apparent only because various products will be produced and because any single product will not account for all the reactant that disappeared.

1. Half-Time

With a first-order or pseudo-first-order reaction, the **half-time,** $t_{1/2,}$ is that period of time required for [S] to reach half its original value of $[S]_0$ (Figure 4-2-A). From Equation 4.7:

$$\log_e 2 = k_{app}\, t_{1/2} \tag{4.11}$$

$$t_{1/2} = \frac{\log_e 2}{k_{app}} = \frac{0.693}{k_{app}} \quad (4.12)$$

The half-time can be used to calculate the apparent rate constant for the reaction using Equation 4.12.

The velocity of a first-order reaction decreases exponentially with time, so it takes a relatively long time to reach completion or equilibrium. After 1–8 half-lives, the extent of a reaction will have reached 50, 75, 87.5, 93.75, 96.83, 98.44, 99.23 and 99.61% of completion.

2. Relaxation Time

A related measure of the rate of a first-order reaction is the **relaxation time**, usually designated τ. It is simply the reciprocal of the apparent first-order rate constant:

$$\tau = \frac{1}{k_{app}} \quad (4.13)$$

After a period of time τ, the concentration of reactant S is at 1/e (= 0.37) its original value, $[S]_0$. The value of τ can also be determined by extrapolating linearly the initial rate to the final extent of the reaction (Figure 4-1). τ is also a measure of the mean duration of the reaction.

3. Reversible Reactions

Most reactions are reversible to some extent and the product can regenerate the reactants, depending upon their relative thermodynamic stabilities. Consequently reactants are rarely converted completely to products, and an equilibrium (Section 1.1) is usually established eventually:

$$S \leftrightarrow P \quad (4.14)$$

If the rate constant for the reverse reaction, k_r, is significant relative to that for the forward reaction, k_f, only some of S is converted to P at equilibrium. In that case, both the forward and reverse rate constants determine the kinetics:

$$\frac{-d[S]}{dt} = k_f[S] - k_r[P] \quad (4.15)$$

At equilibrium, the concentrations of S and P do not change, and the velocity of the forward reaction is equal to that of the reverse reaction:

$$k_f [S]_{eq} = k_r [P]_{eq} \quad (4.16)$$

$$\frac{k_f}{k_r} = \frac{[P]_{eq}}{[S]_{eq}} = K_{eq} \quad (4.17)$$

The constant K_{eq} is the **equilibrium constant** for the reaction and is a function of the difference in the standard free energy difference, $\Delta G°$, between the reactant(s) and product(s) (Section 1.2). A large negative $\Delta G°$ is reflected in a reaction that goes nearly to completion, while $\Delta G° = 0$ will produce an equilibrium in which half the reactant S is converted to P. If $\Delta G° > 0$, less than half of S will be converted to P.

With a reversible reaction, the initial velocity is given by the forward rate constant, but the net velocity decreases as the reaction proceeds, not only because of the decrease in the concentration of S, but also because of the accumulation of the product, P (Figure 4-1-B). In this case, a plot of log[S] versus time is not linear. But a linear plot results if the difference from the equilibrium situation, $[S] - [S]_{eq}$, is used instead (Figure 4-2-A). For this analysis, it is vital to know the value of $[S]_{eq}$ accurately. The apparent rate constant indicated by the negative slope of the linear plot is then the sum of the forward and reverse rate constants:

$$\log_e \left(\frac{[S]-[S]_{eq}}{[S]_0-[S]_{eq}}\right) = -(k_f + k_r)t \tag{4.18}$$

An increased value of the rate constant for the reverse reaction, k_r, increases the apparent rate constant for the forward reaction, but this does not imply that the reaction proceeds more rapidly if it is more reversible. The initial rate of the reaction is determined only by k_f, but the apparent rate constant is greater because the reaction proceeds to a lesser extent, decreasing with the magnitude of the reverse rate constant (Figure 4-1).

4.1.B. Second-Order Kinetics

A reaction between two molecules of S or between S and another molecule, A, follows second-order kinetics. In the first case:

$$2\,S \rightarrow P \tag{4.19}$$

The rate equation is:

$$\text{velocity} = \frac{-d[S]}{dt} = \frac{1}{2}\frac{d[P]}{dt} = k_{bi}\,[S]^2 \tag{4.20}$$

Integration of Equation 4.20 predicts that the time–course of this reaction is described by:

$$\frac{[S]_0}{[S]} = 1 + [S]_0\,k_{bi}\,t \tag{4.21}$$

and the half-time of the reaction is given by:

$$t_{1/2} = (k_{bi}\,[S]_0)^{-1} \tag{4.22}$$

Whether a reaction is first- or second-order in a reactant can be determined by plotting both log[S] and $[S]^{-1}$ versus t and observing which is a straight line. Furthermore, the half-time for a second-order reaction should depend upon $[S]_0$, whereas that for a first-order reaction should not.

In the second-type of second-order reaction, with two different reactants:

$$S + A \rightarrow P \tag{4.23}$$

the rate equation is:

$$\text{velocity} = \frac{-d[S]}{dt} = \frac{-d[A]}{dt} = k_{bi}\,[S]\,[A] \tag{4.24}$$

As the reaction involves two different reactants, their concentrations can be varied independently. If A and S are present at equal concentrations, the rate equation is like that of the first type of second-order reaction (Equation 4.21). If one reactant is present in much greater concentration than the other, so that its concentration remains virtually constant throughout the reaction, the reaction is pseudo-first-order in the limiting reactant. If the limiting reactant is S:

$$\text{velocity} = \frac{-d[S]}{dt} = \frac{-d[A]}{dt} = k_{app}\,[S] \tag{4.25}$$

k_{app} is the pseudo-first-order rate constant, and its value is proportional to the concentration of A. The second-order rate constant can be calculated using the concentration of A:

$$k_{bi} = \frac{k_{app}}{[A]} \tag{4.26}$$

Other situations, in which the concentrations of S and A are not equal but both change during the reaction, yield variable time–courses intermediate between those described by Equations 4.7 and 4.21.

Reactions between two molecules cannot occur more rapidly than the rates at which they encounter each other by diffusion, so second-order rate constants in solution cannot exceed about 10^{10} M^{-1} s^{-1}. In dilute solution, simultaneous encounters between more than two reactants are exceedingly unlikely, so reactions involving more than two reactants usually occur in several steps, with no more than two reactants involved in each.

Kinetic studies of protein–protein interactions. G. Schreiber (2002) *Curr. Opinion Struct. Biol.* **12**, 41–47.

Modeling the kinetics of bimolecular reactions. A. Ferandez-Ramos *et al.* (2006) *Chem. Rev.* **106**, 4518–1584.

4.1.C. Zero-Order Kinetics

A reaction is zero-order in S when the rate of its disappearance is independent of its concentration. Any number raised to the zero power is one, so the term $[S]^0$ would be absent from Equation 4.4:

$$\text{velocity} = \frac{-d[S]}{dt} = k_{app} \tag{4.27}$$

Such kinetics can be observed even with multi-reactant reactions if this particular reactant is not involved in the rate-determining step or if it has been converted to another form prior to the rate-determining step.

4.1.D. Transition State

How far and how fast a reaction will occur are independent variables. The difference in free energy between the reactant and the product determines the equilibrium constant for the reaction (Equation 4.17) but a large and favorable free energy difference does not ensure that the reaction will occur rapidly or at all. The rate of a reaction is determined by the **free energy of the transition state** (or **activated state**), which is an independent variable. The transition state is the least stable species, with the greatest free energy, that occurs during the reaction. The overall reaction will occur rapidly if the transition state has a low free energy; **the higher the free energy of the transition state, the slower the reaction**. Within the transition state for chemical reactions, covalent bonds are being broken and made, so transition states are unstable and often have high free energies. The transition state is usually indicated by the symbol ‡.

Transition state theory was devised for simple chemical reactions involving breakage and formation of covalent bonds in the gas phase. In this most simple case, the observed rate constant for a reaction, k_{app}, is defined by the free energy of the transition state relative to that of the ground state of the reactant, $\Delta G^{\ddagger}$:

$$k_{app} = \left(\frac{k_B T}{h}\right) \exp\left(-\Delta G^{\ddagger}/RT\right) \tag{4.28}$$

k_B is the Boltzmann constant, T the absolute temperature, h the Planck constant and R the gas constant. This equation often includes a **transmission coefficient**, which is the probability that the transition state will break down to products, rather than back to reactants. It is believed usually to have a value between 0.5 and 1.0; unless its value is known, it is usually assumed to be unity.

The transition state $S^{\ddagger}$ is assumed to be in rapid equilibrium with the reactant S:

$$S \underset{}{\overset{K_{eq}^{\ddagger}}{\rightleftharpoons}} S^{\ddagger} \longrightarrow P \tag{4.29}$$

$$[S^{\ddagger}]/[S] = K_{eq}^{\ddagger} = \exp\left(-\Delta G^{\ddagger}/RT\right) \tag{4.30}$$

Moreover, molecules that reach the transition state are assumed to be converted to products by the most rapid process in which covalent bonds can be broken and made in a vacuum, which is given by

($\frac{k_B T}{h}$) and has the value 6×10^{12} s^{-1} at 25° C. The kinetic equation for the reaction S → P is then:

$$\frac{-d[S]}{dt} = [S]^{\ddagger}\frac{k_B T}{h} = [S]\frac{k_B T}{h}\exp(-\Delta G^{\ddagger}/RT) \tag{4.31}$$

which produces Equation 4.28. Consequently, a first-order rate constant k_{app} is sufficient to define the free energy of the transition state relative to that of the reactant for a unimolecular reaction:

$$\Delta G^{\ddagger} = RT \log_e\left(\frac{k_B T}{k_{app} h}\right) \tag{4.32}$$

At 25° C, with k_{app} expressed in s^{-1}, this equation has the form:

$$\Delta G^{\ddagger} = (17.4 - 1.36 \log_{10} k_{app}) \text{ kcal/mol} \tag{4.33}$$

If k_{app} = 1 s^{-1}, $\Delta G^{\ddagger}$ = 17.4 kcal/mol (=72.7 kJ/mol).

The transition state can be considered as a free energy barrier that must be surmounted by fluctuations of the molecule depending upon its energy content. The energy landscapes that control the rates and equilibria of reactions can usefully be compared to mountain landscapes, with the altitude proportional to the free energy and the paths following the easiest pathways between the valleys corresponding to the energies of the pathways that a chemical reaction will follow. The transition state corresponds to the highest pass over the mountain using the most practical path. Other paths would encounter higher peaks to cross, but both hikers and molecules sensibly use the path with the easiest crossing of the peaks.

The transition state also applies to the reverse reaction, so the rate constants for the forward and reverse reactions specify the free energy of the reacting molecule throughout the chemical reaction. The free energy of a reaction is usually described by a **reaction coordinate** (Figure 4-4). The transition state is usually placed midway along the reaction coordinate, unless it is known from other information to be more like the reactant or the product.

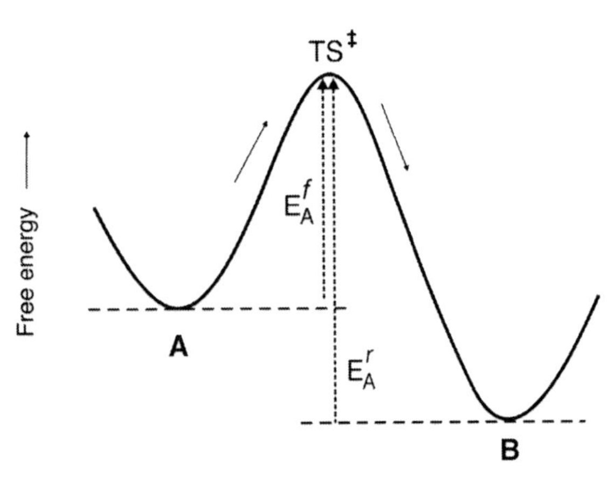

Figure 4-4. An example of a reaction coordinate diagram for a one-step chemical reaction, A ↔ B. The reaction coordinate is a measure of the extent to which the chemical reaction has occurred in a molecule; the starting molecule is on the *left*, the product on the *right*. The free energy of the molecule is given by the *solid line*. The species with the highest free energy is the transition state (TS‡) for the reaction. The higher its free energy, the slower the reaction; the size of the free energy barrier for the forward reaction is given by E_A^f, while that for the reverse reaction is given by E_A^r. The difference in free energies of the reactant and product determines the equilibrium constant for the reaction (Equation 4.17). The transition state is typically placed midway along the reaction coordinate, unless it is known from other information to be more similar to the reactant or the product.

Being the least stable species during a reaction, transition states are populated to the smallest extent and for the least amount of time (< 10^{-12} s), so they cannot be observed directly. Their natures can only be inferred by studying the effects of systematically varying the conditions of the reaction or the structures of the reactants and examining their effects on the rate of the reaction, i.e. on the free energy of the transition state (Section 4.1.E). For example, a polar solvent will usually stabilize a transition state that has more charge separation than the reactants, and thereby increase the rate of the reaction. In contrast, a transition state with less charge separation will be destabilized, and the reaction will be slowed. A nonpolar solvent will have the opposite effects. For example, the reaction:

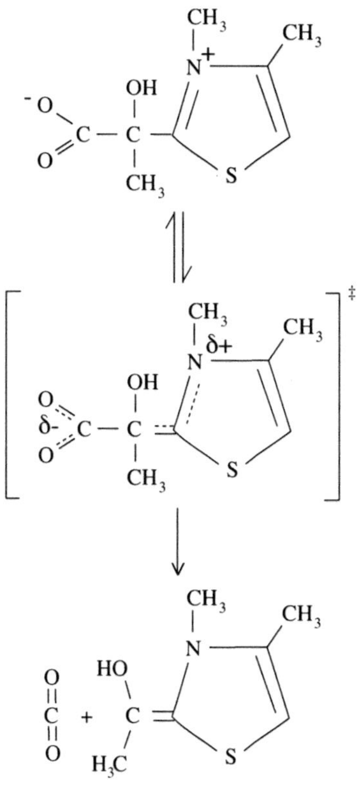

(4.34)

occurs 10^4–10^5 times more rapidly in ethanol than in water. The probable transition state is indicated in brackets. It has the charge separation of the original reactant greatly diminished, so it is stabilized relative to that reactant by a nonpolar solvent and the reaction occurs more rapidly. In contrast, water stabilizes the reactant more than the transition state, thereby decreasing the rate of the reaction. Catalysts, such as enzymes, generally increase the rates of reactions by stabilizing the transition state and lowering its free energy. Transition states in protein folding are being characterized by varying the structure of the protein and measuring the effect on the rate of the folding process.

According to basic transition state theory, **a reaction with no free energy barrier should occur with a rate constant of 6×10^{12} s^{-1} at 25° C.** This assumption is considered reasonable for transition states involving covalent bond breakage and formation in small molecules in the gas phase, but what the barrier-free rate would be in solution or in a complex process like protein folding is a matter for speculation. This difficulty can be avoided by considering not the absolute value of the energy of the transition state relative to the reactant, but only changes in it, for example upon modifying gradually the reactants or the conditions.

The transition state is defined by its increased free energy over that of the reactants, but it can also have, at least in principle, an enthalpy ($H^{\ddagger}$), entropy ($S^{\ddagger}$), and heat capacity ($C_p^{\ddagger}$) (Chapter 1):

$$G^{\ddagger} = H^{\ddagger} - T\,S^{\ddagger} \tag{4.35}$$

$$C_p^{\ddagger} = dH^{\ddagger}/dT \tag{4.36}$$

The corresponding equations for normal ground states are Equations 1.27 and 1.44. The enthalpy of the transition state can be determined from the temperature-dependence of the rate constant. Just as the enthalpy change for a reaction can be obtained from the slope of a plot of log K_{eq} versus temperature^{-1} in a **van't Hoff plot** (Equation 1.31, Figure 1-1), so the value of $\Delta H^{\ddagger}$ can be obtained by plotting log k_{app} versus temperature^{-1}, known as an **Arrhenius plot** (Figure 4-5):

$$\log_e k_{app} = (-\Delta H^{\ddagger}/RT) + (\Delta S^{\ddagger}/R) \tag{4.37}$$

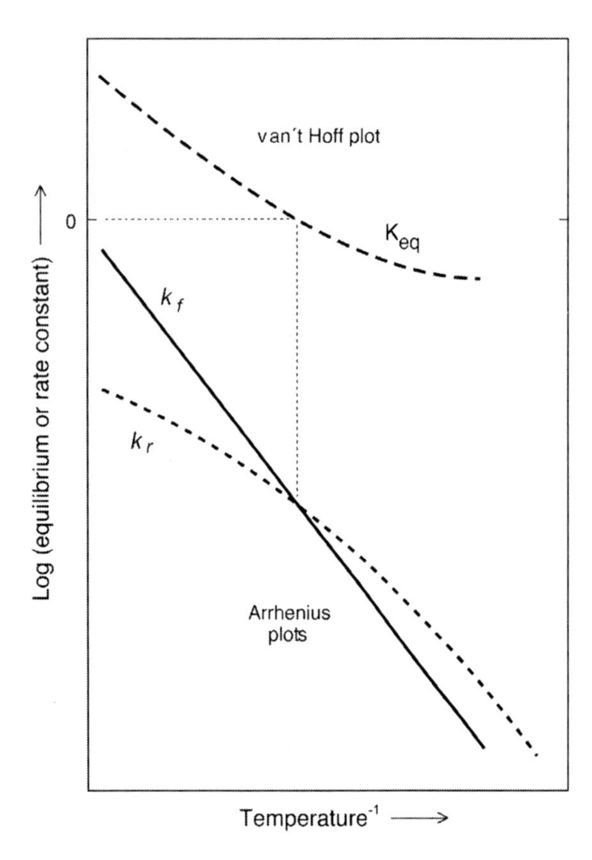

Figure 4-5. Van't Hoff and Arrehnius plots for a hypothetical simple reaction A ↔ B. The curved van't Hoff plot for the equilibrium constant K_{eq} (*top*) indicates that reactant A has a greater heat capacity than the product B. The Arrhenius plot for the rate constant for the forward reaction (k_f) is linear, which indicates that the transition state has the same heat capacity as the reactant A, while that for the reverse reaction (k_r) is curved, indicating that the transition state has a greater enthalpy than the reactant. Even more drastic results than these are observed in the kinetics of protein unfolding and refolding. Note that $K_{eq} = 1$ when $k_f = k_r$.

The slope is equal to $-\Delta H^{\ddagger}/R$; frequently this parameter is referred to as the **activation energy** and can be thought of as the minimum amount of internal energy the molecule requires to initiate the chemical reaction. When the rate of the reaction increases with increasing temperature, the transition state has an enthalpy greater than that of the reactant. Most biochemical reactions under physiological conditions increase in rate approximately 2-fold with an increase in temperature of 10° C. This corresponds to $\Delta H^{\ddagger} = 13$ kcal/mol (54 kJ/mol).

If the transition state has a heat capacity that differs from that of the reactant, $\Delta H^{\ddagger}$ will vary with temperature and the Arrhenius plot will be curved. If there is a change in the heat capacity for the overall reaction, the reactants and products have different heat capacities, so the transition state must also differ from either one or both of the reactants and products. In that case, the Arrhenius plot in at least one direction must be nonlinear. In the hypothetical example shown in Figure 4-5, the Arrhenius plot for the forward reaction is linear, whereas that for the reverse reaction is curved, so the transition state has the same heat capacity as the reactant, A. Even more drastically curved Arrhenius and van't Hoff plots are observed experimentally with complex transitions like protein unfolding and refolding. With a very large change in enthalpy and heat capacity, so that the equilibrium constant goes to zero at the melting temperature, the curvature of the Arrhenius plot for protein refolding causes the rate of refolding to decrease with increasing temperatures, just the opposite of normal chemical reactions.

From the enthalpy and free energy of the transition state, the entropy may be calculated by difference (Equation 4.35). All such detailed properties of the transition state will be valid, however, only if the transition state is truly in rapid equilibrium with the reactant, which is virtually impossible to demonstrate experimentally.

A further limitation of the transition state theory just described is that it does not account for interactions of the reactants with the solvent, nor the dependence of the rates of diffusion on its viscosity. More detailed treatments of reaction rates can include such effects, and that originating from **Kramers** has become of interest for studies on macromolecules. It considers the velocity of a particle starting in its ground state crossing the energy barrier of the transition state, in terms of the mass (m) and shape (as reflected in f, the frictional coefficient) of the reactant, plus the viscosity of the solvent, η_0). The outcome of these calculations is that the frequency N with which a molecule crosses the energy barrier should be given by:

$$N = (\omega_A\ \omega_{\ddagger}/2\pi\ \beta)\ \exp\left(-m\ V_{\ddagger}/k_B\ T\right) \tag{4.38}$$

where ω_A and $\omega_{\ddagger}$ are the angular frequencies of harmonic oscillators when the molecule is in its normal state and transition state, respectively; $V_{\ddagger}$ is the height of the energy barrier, and $\beta = f\eta_0/m$, which is inversely proportional to the sedimentation coefficient of the molecule under the same conditions. Kramers' theory has the advantage of taking into account the effects of the mass and shape of the molecule, plus the viscosity of the solvent, but parameters such as ω_A and $\omega_{\ddagger}$ are not readily defined.

Determining the geometries of transition states by use of antihydrophobic additives in water. R. Breslow (2004) *Acc. Chem. Res.* **37**, 471–478.

Recrossings and transition-state theory. H. O. Pritchard (2005) *J. Phys. Chem. A* **109**, 1400–1404.

Definability of no-return transition states in the high-energy regime above the reaction threshold. C. B. Li *et al.* (2006) *Phys. Rev. Lett.* **97**, 028302.

Skirting the transition state, a new paradigm in reaction rate theory. J. M. Bowman (2006) *Proc. Natl. Acad. Sci. USA* **103**, 16061–16062.

4.1.E. Free-Energy Relationships

The most powerful method of characterizing the transition state for a reaction is to use a series of closely related reactants in the same chemical reaction and to compare the rates of their reaction with their physical properties. For example, the reacting groups in hydrolytic reactions are generally either **nucleophiles**, which have a tendency to donate an electron pair and usually are bases, or **electrophiles**, which tend to accept an electron pair and usually are acids. A nucleophile is converted to an electrophile by protonation: an $-NH_2$ group is a nucleophile, while its protonated $-NH_3^+$ form is an electrophile. The relative strengths of their nucleophilicity and electrophilicity are reflected in their tendencies to acquire or donate protons, which is measured by their pK_a value (Section 3.4). The reaction of a nucleophile with any group would be expected to be similar to its reaction with a proton, so the reactivity of a nucleophile in a chemical reaction is usually related to its pK_a. This is usually apparent by a linear **Brønsted plot**, in which the logarithms of the rate constants for a particular type of chemical reaction, k, involving various nucleophiles or electrophiles are plotted versus their pK_a values. A linear free energy relationship is apparent if the data fall on a straight line according to the general equation:

$$\log k = \beta\, pK_a + \text{constant} \tag{4.39}$$

The slope of the straight line gives the value of the parameter β in the case of nucleophiles, but it is usually designated α in the case of electrophiles. The values of β are usually positive, meaning the nucleophiles with higher pK_a values, and increased tendencies to ionize by binding a proton, react more rapidly, whereas those of α are usually negative, just the opposite.

The reaction used must be the same in each case and involve a series of either nucleophiles with different pK_a values attacking the same ester or amide bond (Figure 4-6) or one nucleophile attacking a series of molecules with various leaving groups having different pK_a values. The reactants must be closely related, and no special factors such as steric hindrance should affect their rate of reaction to varying extents. In such cases, linear correlations are found frequently.

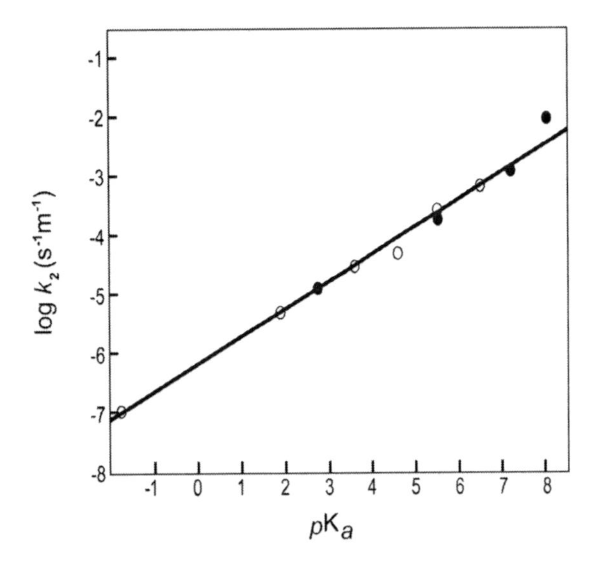

Figure 4-6. Example of a typical Brønsted plot for the general-base catalysis of the hydrolysis of an ester. The logarithms of the second-order rate constants measured are plotted against the pK_a of the catalytic base. The slope gives the value of β. The *open circles* are for amine bases, while the *closed circles* are for oxyanion bases. Their similar effects indicate that the catalysis depends primarily on the basic strength of the base and not on its chemical nature. Data from A. Fersht (1985) *Enzyme Structure and Mechanism*, 2nd edn, W. H. Freeman, NY.

The value of α or β provides important information about the transition state for the reaction. Its magnitude is often interpreted as indicating the amount of charge developed in the transition state and whether the physical properties of the transition state are closer to the reactants or the products. The absolute values of α and β generally fall between 0 and 1; a value of 1 would indicate complete transfer of a proton in the transition state, while a value of 0 indicates no transfer. Absolute values between 0.3 and 0.6 are usually measured for ester hydrolysis reactions (Figure 4-6); positive values are observed for series of attacking nucleophiles, whereas negative values are observed with variation of the alcohol leaving group. On the other hand, very small absolute values of 0.1–0.2 are observed with the reactions of very basic nucleophiles with reactive esters (e.g. phenolates), indicating that the transition state in this case is closer to the initial reactants. In contrast, much larger values, close to 1.5, are observed for the attack of bulky tertiary amines on esters that have poor leaving groups; such a very large value suggests that the transition state for this reaction is close to the products.

Free energy relationships were developed with simple chemical reactions between small molecules, but they have also been found to be useful in characterizing the transition states in phenomena as complex as protein folding. In these cases, the rate constant is compared with the equilibrium constant for the reaction, to determine whether the transition state is more similar to the initial state or to the final state.

Non-linear rate-equilibrium free energy relationships and Hammond behavior in protein folding. I. E. Sanchez & T. Kiefhaber (2003) *Biophys. Chem.* **100**, 397–407.

Relationship of Leffler (Bronsted) α values and protein folding Φ values to position of transition-state structures on reaction coordinates. A. R. Fersht (2004) *Proc. Natl. Acad. Sci. USA* **101**, 14338–14342.

From shut to open: what can we learn from linear free energy relationships? D. Colquhoun (2005) *Biophys. J.* **89**, 3673–3675.

4.2. MULTI-STEP REACTIONS AND INTERMEDIATES

Chemical reactions in dilute solutions that involve more than two reactants in each step are very rare, as **collisions of more than two molecules simultaneously are very unlikely**. Consequently, any chemical reaction involving a number of reactants, as in Equation 4.4, is likely to occur in several sequential steps:

$$S \rightarrow A \rightarrow B \rightarrow C \rightarrow P \tag{4.40}$$

with at most two of the reactants being involved in each step; S is the reactant, P the final product, and A, B and C are intermediates. Any steps that involve another reactant will be second-order. Kinetic analysis of such a sequence of steps is more complex than with a single step, and a full kinetic description will involve describing the rate constants for the appearance and disappearance of all the reactants, including the intermediates. Complex reaction schemes usually do not yield analytical expressions readily, and complete analysis of their time–course is best carried out by numerical integration of the kinetic equations.

The intermediates will be formed in order of their occurrence in the reaction and will then disappear with time (Figure 4-7). Each of these intermediates will be present at significant concentrations if all of the steps occur at similar rates. Also, the product P will appear only if its immediate precursor C is present, so its initial rate of appearance will be zero and will then gradually increase as the intermediate C is generated. Consequently, there will be a **lag period** in the appearance of the later intermediates and of the product. **The greater the number of intermediates, the greater the lag period in appearance of the final product** (Figure 4-8). The lag period is longest when all the rate constants are identical, when all the steps limit the rate of the reaction.

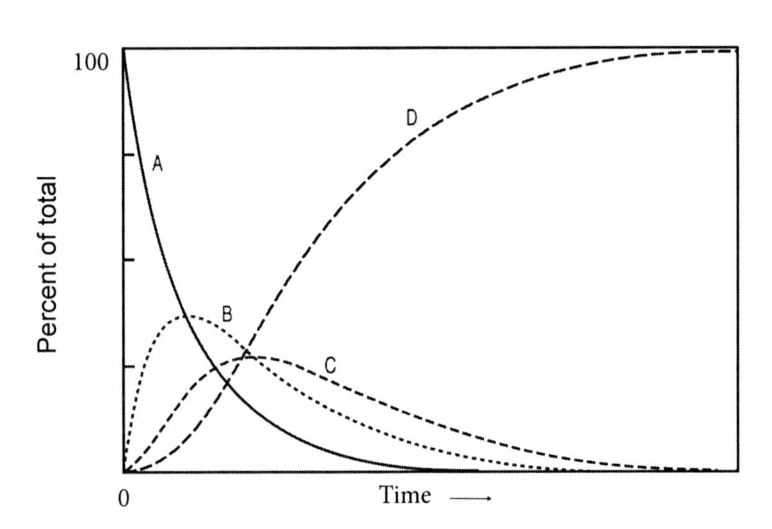

Figure 4-7. Time-course of an obligatorily sequential pathway with multiple rate-limiting steps. The progress of the irreversible reaction A → B → C → D was simulated with each step having the same unimolecular rate constant. The proportion of molecules in each species is plotted as a function of time. Note that A disappears and B appears without a lag, but that C and D appear only after lag periods. The lag period for C corresponds to the time during which the concentration of its precursor B is increasing. The lag period in formation of D is longer due to the need to build up the concentration of C as well. Adapted from T. E. Creighton (1990) Biochem. J. 270, 1–16.

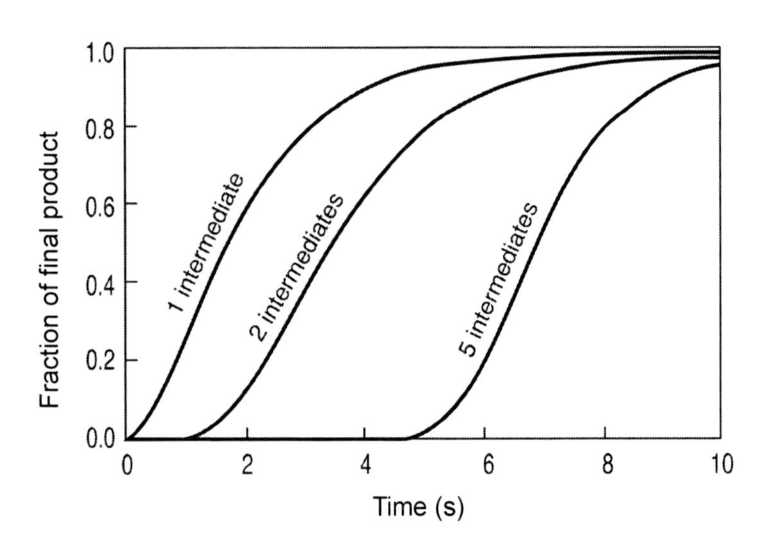

Figure 4-8. Increasing lag period in the appearance of the final product of a multi-step reaction involving different numbers of intermediates. The time-course of appearance of the final product is illustrated. In each case, each step in the reaction sequence has the same rate constant ($1\ s^{-1}$). During the lag period, the concentrations of the various obligatory intermediates are increasing. The reaction for the curve with one intermediate has two kinetic steps, that with two intermediates has three, and that with five intermediates has six steps. Data from H. Gutfreund (1995) *Kinetics for the Life Sciences*, Cambridge University Press, Cambridge, p. 121.

The free-energy profile of a multi-step reaction may be constructed from the rate and equilibrium constants for all of the steps. This is straightforward for a solely unimolecular reaction (Figure 4-9) but the rate and equilibrium constants will depend upon the concentrations of any other reactants involved in bimolecular steps.

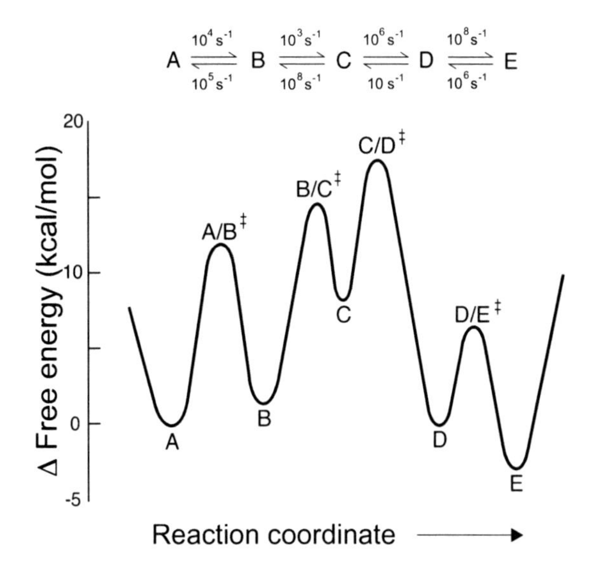

Figure 4-9. Free energy profile for a hypothetical multi-step reaction A ↔ E that proceeds sequentially through intermediates B, C and D. The rate constants for all the steps are given at the top; they specify the relative free energies of all the species and all the transition states involved in their interconversion. The overall transition state that determines the rates of the reaction in both directions is that with the highest free energy, between intermediates C and D. Adapted from T. E. Creighton (1993) *Proteins: structures and molecular properties*, 2nd edn, W. H. Freeman, NY, p. 392.

Sequential vs. parallel protein-folding mechanisms: experimental tests for complex folding reactions. L. A. Wallace & C. R. Matthews (2002) *Biophys. Chem.* **101**, 113–131.

Reaction progress kinetic analysis: a powerful methodology for mechanistic studies of complex catalytic reactions. D. G. Blackmond (2005) *Angew. Chem. Int. Ed. Engl.* **44**, 4302–4320.

4.2.A. Rate-Determining Step

One of the steps in a multi-reaction sequence is likely to be slower than the others, and its rate will determine the rate of the overall reaction; this is known as the **rate-determining** or **rate-limiting** step. The rate equation of the overall reaction will be given by that for this step. For example, for a two-step reaction such as:

$$A + B \rightarrow C \tag{4.41}$$

$$C + D \rightarrow P \tag{4.42}$$

the second step might occur much more rapidly than the first, so the first step would be rate-limiting. The rate equation for the overall reaction would then be:

$$\text{velocity} = k_{app}\,[A]\,[B] \tag{4.43}$$

The concentration of reactant D would not appear in the rate equation and would not govern the rate of the reaction, so long as it was sufficient for the second reaction to be much faster than the first. The overall reaction would be zero-order in D. **Intermediates that occur after the rate-limiting step will not accumulate to detectable levels**, as they are rapidly converted to the final product.

Alternatively, the second step might be rate-limiting, so the rate equation would be:

$$\text{velocity} = \frac{d[P]}{dt} = k_{app}\,[C][D] \tag{4.44}$$

C is not an initial reactant, however, so the rate equation will depend upon how C is formed from A and B. Intermediate C might be in rapid equilibrium with A and B during the reaction, with equilibrium constant K_{eq}. The overall rate equation would then be given by:

$$\text{rate} = \frac{d[P]}{dt} = k_{app}\,K_{eq}\,[A][B][D] \tag{4.45}$$

The extent to which C accumulates will depend upon its free energy relative to A and B. Note that this is a third-order reaction, but that no more than two molecules are involved in each step.

For an irreversible reaction sequence of n steps, like that of Equation 4.40, the apparent rate constant, k_{app}, and relaxation time, τ_{app} (after the lag period) will be given by:

$$\frac{1}{k_{app}} = \frac{1}{k_1} + \frac{1}{k_2} + \frac{1}{k_3} + \ldots \tag{4.46}$$

$$\tau_{app} = \tau_1 + \tau_2 + \tau_3 + \ldots \tag{4.47}$$

where k_i is the first-order or pseudo-first-order rate constant for step i and τ_i is the corresponding relaxation time (Section 4.1.A.2). This equation illustrates how the slowest step predominates in determining the overall reaction rate, and that multiple slow steps decrease the observed rate. Where one or more steps are reversible, the reaction is best characterized in the steady state (Section 4.2.B).

The rate-determining step is apparent from the free energy profile of the overall reaction, such as that in Figure 4-9. It is the step with the transition state with the highest free energy overall. In that example, the rate-limiting step for the overall conversion of A to E would be C → D, even though the smallest individual rate constant is the one that produces C. Intermediate C is very unstable and is rapidly reversed to B. The rate-determining transition state is that between C and D, and the rate constant for the overall reaction will be determined by the free energy of this transition state relative to the original reactant, A. In this example, only intermediate B would accumulate transiently to significant levels during the reaction A → E, but as no more than 10% of the molecules. Intermediate D is more stable than A, but it would not accumulate to significant levels because it occurs after the rate-determining step and would be converted very rapidly to P.

The situation is more complex if any of the steps are second-order and involve other reactants, because then the apparent free energies of the various intermediates and transition states will depend upon the concentrations of the other reactants. Consequently, changing the concentrations of the reactants can change which step is rate-determining. This can lead to complex kinetics that are of mixed-order.

As an example, the conversion of an aldehyde, R_1–CHO, into R_1–CH=N–R_2 occurs by the aldehyde reacting first with an amine, H_2N–R_2, to form a carbinolamine intermediate, which then loses a water molecule:

$$R_1\text{–CHO} + H_2N\text{–}R_2 \leftrightarrow R_1\text{–CH(OH)NH–}R_2 \rightarrow R_1\text{–CH=N–}R_2 + H_2O \tag{4.48}$$

The first step is reversible, with rate constants k_1 and k_{-1} and equilibrium constant $K_{eq} = k_{-1}/k_1$; the rate constant for the second step is k_2. At low concentrations of amine and pH, the rate-determining step is the formation of the carbinolamine intermediate. At high concentrations of amine and neutral pH, however, the intermediate is formed completely, and the rate of the reaction no longer depends on the concentration of the amine; the kinetics are zero-order with respect to it. The rate-determining step has changed, and the overall order is mixed. If the concentration of the amine is greater than 10 K_{eq}, the reaction will become pseudo-first-order:

$$\frac{-\,d[R_1\text{–CHO}]}{dt} = k_{app}[R_1\text{-CHO}] \tag{4.49}$$

$$k_{app} = \frac{k_2[H_2N - R_2]}{K_{ea}} \tag{4.50}$$

The same step is usually rate-determining in the reverse direction, so long as the conditions are the same, a principle known as **microscopic reversibility**. This is apparent by regarding Figure 4-9 for the reverse reaction. Reactants that occur after the rate-determining step in the forward reaction do not affect its rate, but they do affect the reverse reaction, because their presence reverses one of the steps preceding the rate-determining step in that direction. For example, the reactant D is involved after the rate-limiting step B → C in the forward direction in the reaction:

$$A \underset{K_1}{\overset{E}{\longleftrightarrow}} B \underset{k_r}{\overset{k_f}{\rightleftharpoons}} C \underset{K_2}{\overset{D}{\longleftrightarrow}} F \tag{4.51}$$

and the rate in that direction is independent of D's concentration (so long as it is sufficient for the last step not to be rate-determining). In contrast, the rate equation for the reverse direction (in the absence of A and B) will be inversely proportional to the concentration of D:

$$-\frac{d[F]}{dt} = k_r[C] = \frac{k_r[F]}{k_2[D]} \tag{4.52}$$

The reverse reaction will be slowed by increasing concentrations of D because it will rapidly reverse the first step in the disappearance of F. Similarly, any product produced prior to the rate-determining step in the other direction, such as E in Equation 4.51, will slow the forward reaction when present at substantial concentrations:

$$-\frac{d[A]}{dt} = k_f[B] = \frac{k_f K_1[A]}{[E]} \tag{4.53}$$

All of these considerations follow simply from the requirement that the ratio of the forward and reverse rate constants be equal to the overall equilibrium constant, $K_{overall} = k_1 k_2 (k_f/k_r)$, and all of these reactants enter into the expression for the equilibrium constant. At equilibrium:

$$\frac{k_r[F]_{eq}}{K_2[D]_{eq}} = k_f K_1 \frac{[A]_{eq}}{[E]_{eq}} \tag{4.54}$$

$$\frac{[E]_{eq}[F]_{eq}}{[A]_{eq}[D]_{eq}} = \frac{k_f}{k_r} K_1 K_2 = K_{overall} \tag{4.55}$$

4.2.B. Steady-State Kinetics

The concentrations of the intermediate species in a multi-step reaction, like A, B and C in Equation 4.40, change relatively little after the lag period, especially when they are present at only low concentrations. They will be generated from S at about the same rate as they are converted to P, and their concentrations will decrease only as that of S diminishes. Such a reaction is in a **steady-state**. The rates of formation of the intermediates are nearly equal to their rates of breakdown, and the situation can be analyzed by setting the rate equations for their appearance equal to those for their disappearance. This is known as the **steady-state approximation**, as it does not apply exactly. It is very useful in analyzing complex reactions.

The steady-state concentration can then be used in the equations for the rate of the overall reaction. For example, in Equation 4.48, the steady-state concentration of the intermediate can be estimated by setting equal its rates of formation and disappearance:

$$k_1 [R_1\text{-CHO}] [H_2N\text{-}R_2] = (k_{-1} + k_2) [R_1\text{-CH(OH)NH-}R_2]_{ss} \tag{4.56}$$

$$[R_1\text{-CH(OH)NH-}R_2]_{ss} = \frac{k_1}{k_{-1} + k_2} [R_1\text{-CHO}][H_2N\text{-}R_2] \tag{4.57}$$

This steady-state concentration can then be inserted into the rate equation for the overall reaction:

$$\text{velocity} = k_2[R_1\text{-CH(OH)NH-}R_2]_{ss} = \frac{k_1 k_2}{k_{-1} + k_2}[R_1\text{-CHO}][H_2N\text{-}R_2] \tag{4.58}$$

Similar manipulations of the scheme in Figure 4-9 indicate that the steady-state concentrations of intermediates A, B and C will be 10^{-1}, 10^{-6} and 10^{-8} the concentration of the reactant S during the forward reaction.

4.3. MEASURING RAPID REACTIONS

Within a multi-step reaction, like that of Equation 4.40, it is necessary to elucidate the kinetic roles of all the intermediate species, to determine whether they are productive or nonproductive intermediates, i.e. on or off the pathway, and to determine their order in the pathway. This can only be done by measuring the kinetics of formation and disappearance of each of the species. Measuring the approach to a steady-state generally requires rapid reaction techniques, as individual reactions can occur on very short time scales.

Advances in transient-state kinetics. K. A. Johnson (1998) *Curr. Opinion Biotechnol.* **9**, 87–89.

4.3.A. Rapid Mixing Techniques

The rates of chemical reactions are generally measured by mixing the reactants at time zero and following the disappearance of the reactants and the appearance of the products in real time. **Monitoring the reaction using spectrophotometric method is most convenient**, and it is usually possible because the various species are likely to differ in some suitable property. Reactions that occur at the minute time scale can usually be initiated by manual mixing. The rates of bimolecular reactions can be controlled by varying the concentrations of the reactants, and rapid reactions can be slowed markedly by decreasing the temperature or altering the pH. Nevertheless, there are many occasions when reactions occur too rapidly to be measured readily using manual mixing of reactants, and various techniques for rapid mixing have been devised. Generally, solutions containing the two reactants are mixed rapidly in a special mixing chamber designed to produce turbulent flow to merge the two solutions rapidly and efficiently; the combined mixture then travels along capillary tubes while the reaction takes place. The greatest technical challenge with such instruments is to ensure rapid mixing and uniform flow of solutions along the capillaries. The flow rate of the liquid through the capillaries must be kept above a critical velocity in order to ensure 'turbulent flow'. Below this velocity, which is about 2 m s^{-1} for a tube with a 1-mm diameter, the flow may be laminar, when the liquid at the center of the tube travels faster than that near the wall. This requirement for rapid flow places an upper limit on the time that the reaction can be retained within such an instrument.

Most simple is the **continuous flow** apparatus, in which the mixed solution flows along a linear capillary, where the extent of the reaction is monitored, usually spectrophotometrically (Figure 4-10). The various positions along this capillary correspond to different times since the two reactants were mixed, which depend upon the flow rate. The position closest to the mixing chamber usually corresponds to times of reaction of about 1 ms (e.g. a position 1 cm past the mixing chamber with a flow rate of 10 m s^{-1}). In this case, it is feasible to monitor kinetic reactions with apparent rate constants approaching 10^3 s^{-1} (relaxation times of 1 ms). The flow rate must be maintained constant throughout the observations along the length of the capillary, so large volumes of reactants must be used, and the longest reaction times practical are about 100 ms. Once the reactants have been exhausted, and the flow stopped, any subsequent reaction can be monitored in real time. Making observations at different points along the capillary, however, presents technical problems.

Requiring less material is the **stopped-flow** apparatus, and it has replaced continuous flow (Figure 4-11). Small amounts of reactants (50–200 μl) are mixed rapidly, and the flow is then stopped quickly. The reaction is monitored in real time at only one position near the mixing chamber, to be able to monitor the earliest stages of the reaction. This 'dead-time' can usually be just one or a few milliseconds, but it is longer than with continuous flow because of the need to stop the flow before measurements can commence. The reaction in the stationary mixture can then be monitored for several minutes.

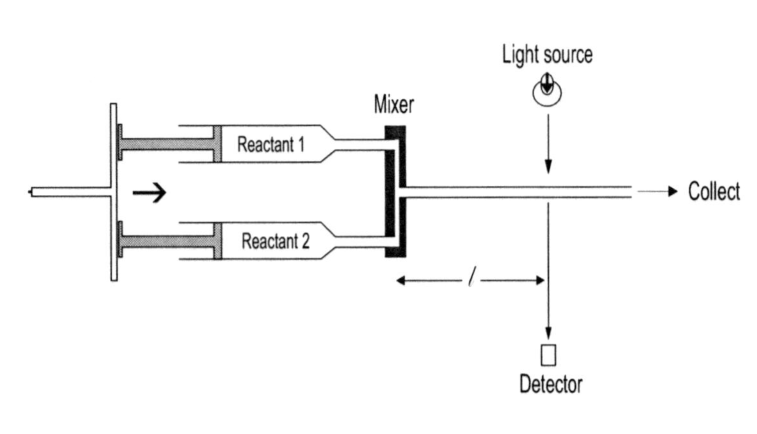

Figure 4-10. Schematic drawing of a continuous-flow apparatus for monitoring the chemical reaction between two reactants. The two reactants are mixed by gradually and uniformly driving the plungers of the two syringes to the right. The progress of the reaction is followed by monitoring it at various positions l along the capillary. The greater l, the longer the reaction has taken place since the time when the reactants were mixed. If the mixed solution is moving at a rate of x m s^{-1}, the time since the two solutions mixed will be given by l/x. Adapted from A. Fersht (1985) *Enzyme Structure and Mechanism*, 2nd edn, W. H. Freeman, NY, p. 122.

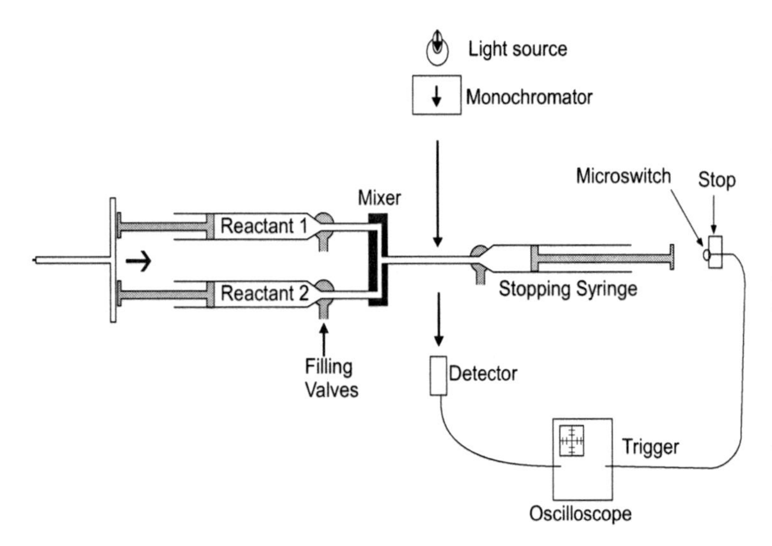

Figure 4-11. Schematic drawing of a stopped-flowapparatusformonitoring the chemical reaction between two reactants. At time zero, small volumes of the two reactants are mixed rapidly, until the stopping syringe is driven to the right and hits the stop position. The progress of the reaction is then monitored optically in real time at a single position. Adapted from A. Fersht (1985) *Enzyme Structure and Mechanism*, 2nd edn, W. H. Freeman, NY, p. 123.

Some reactions cannot be monitored spectrophotometrically, but the various reactants, intermediates and products must be analyzed by chromatographic, electrophoretic or other separation methods. This is possible if the reaction can be quenched in some way, such as by adding an excess of acid to lower the pH very rapidly, to slow the reaction or to destroy one of the reactants. In a **rapid quenching** apparatus (Figure 4-12), the quenching solution is combined with the reactants at the desired time after the reaction is initiated, and the quenched solution is collected and analyzed. Much more sophisticated apparatuses than that illustrated in Figure 4-12 are possible, providing more control over the time and manner of adding the quenching solution.

Other techniques quench the reaction by very **rapid freezing**.

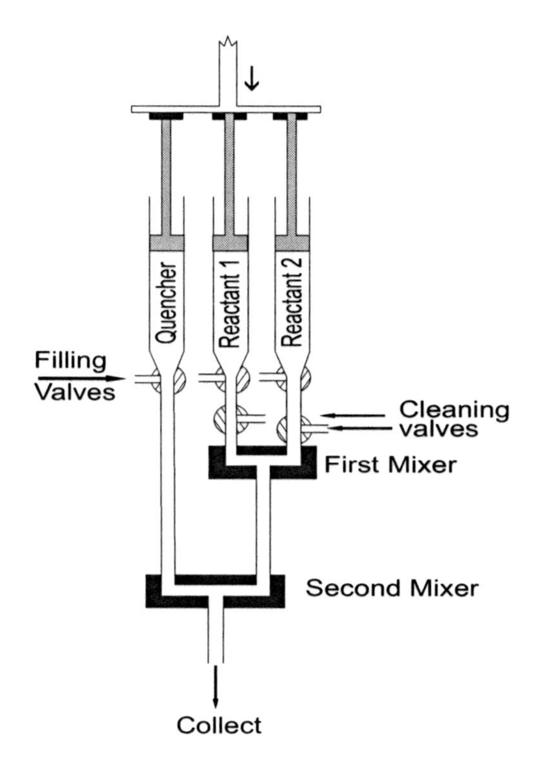

Figure 4-12. Schematic drawing of a quenched-flow apparatus for monitoring the reaction between two reactants. The two reactants are mixed by depressing the plungers of the three syringes simultaneously. The two reactants are mixed, and the reaction between them takes place until the reaction mixture is mixed with the quenching solution. The effluent is collected and analyzed to measure the extent of the reaction. Adapted from A. Fersht (1985) *Enzyme Structure and Mechanism*, 2nd edn, W. H. Freeman, NY, p. 124.

Microsecond freeze-hyperquenching: development of a new ultrafast micro-mixing and sampling technology and application to enzyme catalysis. A. V. Cherepanov & S. DeVries (2004) *Biochim. Biophys. Acta* **1656**, 1–31.

Rapid mixing methods for exploring the kinetics of protein folding. H. Roder *et al.* (2004) *Methods* **34**, 15–27.

The identification of chemical intermediates in enzyme catalysis by the rapid quench-flow technique. T. E. Barman *et al.* (2006) *Cell. Mol. Life Sci.* **63**, 2571–2583.

4.3.B. Relaxation Techniques

Very rapid reactions can be measured by perturbing a system that is at equilibrium and then monitoring the change to the new equilibrium position (Figure 4-13). For example, the temperature or pressure of a solution can be altered very rapidly, within 0.1–1 μs, sufficient to perturb an existing equilibrium by a small amount. The approach to the new equilibrium can be followed at the microsecond time scale, which allows the measurement of kinetic rate constants as large as 10^6 s^{-1}. The equilibrium studied can be a normal reversible chemical reaction or the association of molecules to form a complex.

Temperature-jump is the method used most commonly. The temperature of an aqueous reaction mixture can be increased by 5 to 10° C in about 1 μs by discharging electrical energy into it. The extent to which an equilibrium is perturbed by the temperature change depends upon its enthalpy change, which must not be zero. The shift to the new equilibrium can be followed spectrophotometrically, using specific absorbance or fluorescent probes in macromolecules; alternatively, changes in ionization can be followed using pH-sensitive dyes.

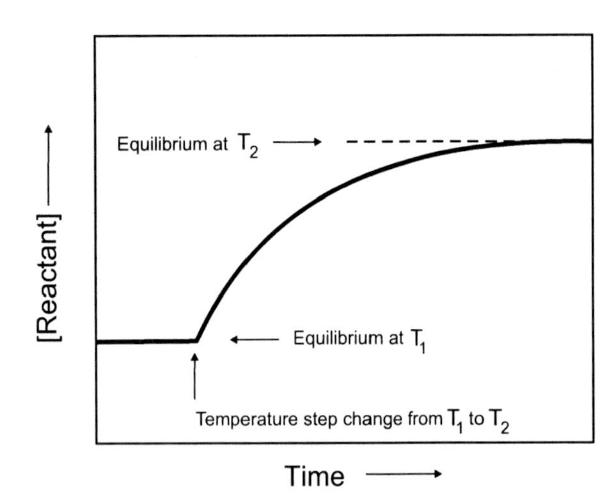

Figure 4-13. Illustration of a temperature jump. A system at equilibrium is subjected to an abrupt change in temperature, and the time-course of attaining the new equilibrium position is followed.

Perturbing the equilibrium by only a small amount simplifies the kinetic analysis, as the concentrations of none of the species change substantially, so each step in a reaction is observed as a single first-order kinetic reaction. Both forward and reverse steps contribute in an equilibrium situation, and the sum of the forward and reverse rate constants governs the apparent rate constant for each step. The apparent rate constants for the various types of reactions are described in Table 4-1.

Table 4-1. Expressions for the reciprocal relaxation times (τ^{-1}) for perturbation of some simple reactions

Reaction	τ^{-1}
$A \underset{k_r}{\overset{k_f}{\rightleftharpoons}} B$	$k_f + k_r$
$A + B \underset{k_r}{\overset{k_f}{\rightleftharpoons}} C$	$k_f([A] + [B]) + k_r$
$A + B \underset{k_r}{\overset{k_f}{\rightleftharpoons}} C + D$	$k_f([A] + [B]) + k_r([C] + [D])$
$2A \underset{k_r}{\overset{k_f}{\rightleftharpoons}} A_2$	$4k_f[A] + k_r$

Pressure-jump relaxation kinetics of a DNA triplex helix-coil equilibrium. M. C. Lin & R. B. Macgregor (1997) *Biopolymers* **42**, 129–132.

Fast kinetics and mechanisms in protein folding. W. A. Eaton *et al.* (2000) *Ann. Rev. Biophys. Biomol. Structure* **29**, 327–359.

The use of pressure-jump relaxation kinetics to study protein folding landscapes. J. Torrent *et al.* (2006) *Biochim. Biophys. Acta* **1764**, 489–496.

~ CHAPTER 5 ~

ISOTOPES AND RADIOACTIVITY

Isotopes have always been of prime importance in molecular biology. Their initial use was to distinguish otherwise identical molecules on the basis of their origin. For example, DNA or proteins synthesized during a brief window of time could be labeled with a radioactive or heavier isotope and thus distinguished from all other molecules of the same type that had been synthesized either before or afterwards. The flexibilities of the structures of proteins and nucleic acids can be measured by following the exchange of their H atoms with different isotopes in the solvent. Now they are also of immense use in nuclear magnetic resonance spectroscopy and vibrational spectroscopy.

Radioisotopes in Biology: a practical approach. R. J. Slater (ed.) (1990) Oxford University Press, Oxford.

Recent advances in the applications of radioisotopes in drug metabolism, toxicology and pharmacokinetics. D. Dalvie (2000) *Curr. Pharm. Des.* **6**, 1009–1028.

5.1. ISOTOPES

Isotopes are atomic species with the same **atomic number** (Z, the number of protons in the nucleus) and therefore belong to the same element, but they have different **mass numbers** (A, the number of protons plus neutrons, known collectively as nucleons) because they have different numbers of neutrons in their nucleus. **Every isotope of a given element has the same number of protons in its atomic nucleus, but a different number of neutrons**. For example, all carbon atoms are distinguished by their atomic number of 6, so they all have six protons in their nucleus, but they can have 2 to 14 neutrons and mass numbers of 8 to 20. Isotopes are distinguished here by their mass number, which is given as a preceding superscript, while the normal chemical symbol for the element is used. If the atomic number is given as well, it is a preceding subscript. Therefore, all carbon atoms have an atomic number of 6, which is implied by the chemical notation 'C' or stated explicitly as '${}_{6}C$'. The various isotopes with mass numbers of 10 to 15 are depicted as ${}^{10}C$, ${}^{11}C$, ${}^{12}C$, ${}^{13}C$, ${}^{14}C$ and ${}^{15}C$. The ${}^{12}C$ isotope predominates naturally, comprising 98.892% of all C atoms, while the isotope ${}^{13}C$ makes up most of the remainder.

The isotopes commonly encountered in molecular biology are listed in Table 5-1. Each of the 110 elements of the periodic table known has more than one isotope, and the total number of known isotopes is more than 1500. Many isotopes do not occur naturally and need to be synthesized.

The various isotopes of an element do not usually differ in their chemical properties because they have the same electronic structure and undergo the same chemical reactions. Only their differences in mass affect their chemical reactivity, and this is usually insignificant, except for the isotopes of hydrogen, deuterium and tritium, because their relative masses (1, 2 and 3) differ three-fold. The varying masses of the isotopes cause shifts in the wavelengths at which they vibrate and absorb radiation, so isotope editing is a useful method of analyzing complex vibrational spectra. The various isotopes can also be distinguished by their varying masses using mass spectrometry (Chapter 6). Different isotopes also differ in their nuclear spins, which are very important for nuclear magnetic resonance.

The masses of atoms are usually expressed relative to the isotope carbon-12, which is defined as having a mass of 12.000000 Daltons (Da, 1 Da = 1 g/mol). Biological macromolecules also contain 1.1% of carbon-13, which has a mass of 13.003355, so the average mass of natural C atoms is 12.011. Similarly, natural molecules contain 0.4% of their nitrogen molecules as nitrogen-15, with a mass of 15.000109, in place of the normal nitrogen-14, with a mass of 14.003074. Very precise mass measurements, as in mass spectrometry (Chapter 6), are affected by the various isotopes present, and the various isotopes can often be distinguished in this way.

Positron-emitting isotopes produced on biomedical cyclotrons. P. McQuade *et al.* (2005) *Curr. Med. Chem.* **12**, 807–818.

Quantitating isotopic molecular labels with accelerator mass spectrometry. J. S. Vogel & A. H. Love (2005) *Methods Enzymol.* **402**, 402–422.

Mass spectrometry and isotopes: a century of research and discussion. H. Budzikiewicz & R. D. Grigsby (2006) *Mass Spectrom. Rev.* **25**, 146–157.

Isotope-edited IR spectroscopy for the study of membrane proteins. I. T. Arkin (2006) *Curr. Opinion Chem. Biol.* **10**, 394–401.

5.2. RADIOACTIVE DECAY

Many isotopes are stable, but the nuclear configurations of **radioisotopes** (or **radionuclides**) are not, and they decay in a spontaneous radioactive transformation to a more stable energy state. This results in the formation of new elements (**decay products**) and the release of energetic particles or photons of energy to compensate for the decreases in their mass.

Radioactivity is one of the physical phenomena used most frequently in molecular biology. **Radioactive isotopes can usually replace their stable, normal counterparts without any substantial change in their chemical properties, yet they can be detected readily by the radiation they emit when they decay.** Consequently, one subpopulation of otherwise identical molecules can be distinguished from another by which isotopes they contain; this is very useful for tracing the fates of molecules in complex biological systems. Proteins and nucleic acids present in only very small quantities can be detected solely on the basis of radioactive isotopes incorporated into them during their biosynthesis. Some

Table 5-1 Isotopes encountered frequently in molecular biology

Element	Atomic number (Z)	Mass number	Isotope	Natural abundance (%)
Hydrogen	1	1	^{1}H	99.9844
		2	^{2}H	0.0156
		3	^{3}H	
Carbon	6	12	^{12}C	98.982
		13	^{13}C	1.108
		14	^{14}C	
Nitrogen	7	14	^{14}N	99.64
		15	^{15}N	0.36
Oxygen	8	16	^{16}O	99.76
		17	^{17}O	0.04
		18	^{18}O	0.20
Sodium	11	23	^{23}Na	100
Magnesium	12	24	^{24}Mg	78.6
		25	^{25}Mg	10.1
		26	^{26}Mg	11.3
Phosphorous	15	31	^{31}P	100
		32	^{32}P	Synthetic
		33	^{33}P	Synthetic
Sulfur	16	32	^{32}S	95.06
		33	^{33}S	0.74
		34	^{34}S	4.18
		35	^{35}S	Synthetic
Chlorine	17	35	^{35}Cl	75.4
		37	^{37}Cl	24.6
Potassium	19	39	^{39}K	93.1
		40	^{40}K	0.0119
		41	^{41}K	6.9
Calcium	20	40	^{40}Ca	96.92
		42	^{42}Ca	0.64
		43	^{43}Ca	0.13
		44	^{44}Ca	2.13
		46	^{46}Ca	0.0032
		48	^{48}Ca	0.179
Manganese	25	55	^{55}Mn	100
Iron	26	54	^{54}Fe	5.90
		56	^{56}Fe	91.52
		57	^{57}Fe	2.245
		58	^{58}Fe	0.33
Copper	29	63	^{63}Cu	69.09
		65	^{65}Cu	30.91
Iodine	53	125	^{125}I	Synthetic
		127	^{127}I	100
		131	^{131}I	Synthetic

molecules can be labeled with radioactive groups so that they can be detected specifically, without interference by the many unlabeled molecules also present. Nucleic acids can be labeled specifically at one end of the polynucleotide chain so that, after some chemical manipulation, only fragments containing that end are detected; this is crucial in sequencing nucleic acids.

Radioactivity takes several different forms, depending upon the structure of the radionuclide's atomic nucleus. **The process of radioactive disintegration (decay) results in changes to the number of nucleons (protons plus neutrons) and to the atomic number of the nucleus, so its chemical identity changes**; in addition, there is release of energetic particles and photons of energy. The particles can be **alpha particles, beta particles** or **positrons**. The photons are **gamma rays** and **X-rays**, plus **neutrinos**, which are discrete packets of energy without mass or charge.

The total energy of the photons and charged particles emitted during radioactive decay accounts for the net decrease in the mass of the disintegrating atom. The energy, momentum and electronic charge of the atom are conserved. The emitted energy can be in the form of either kinetic energy of the particles in motion or quantum energy of photons, both of which ultimately degrade into heat. The energy emitted in radioactive decay is usually measured in **millions of electron volts** (MeV). One electron volt is the energy acquired by any charged particle carrying unit electronic charge when it falls through a potential difference of one volt (1 eV = 3.8×10^{-20} cal or 1.6×10^{-19} J).

Radioisotopes may decay to either stable or other radioactive species. Decay from one radioisotope to another is called a **decay series** or **serial transformation**, such as the radioactive decay of krypton-90:

$$^{90}_{36}\text{Kr} \xrightarrow{\beta^-} {}^{90}_{37}\text{Rb} \xrightarrow{\beta^-} {}^{90}_{38}\text{Sr} \xrightarrow{\beta^-} {}^{90}_{39}\text{Y} \xrightarrow{\beta^-} {}^{90}_{40}\text{Zr} \tag{5.1}$$

In each step, one neutron is converted to a proton. Each step follows first-order kinetics, and the entire series behaves as explained for complex sequential reactions in Section 4.2.

All radioactive substances are health hazards to various degrees, because the particles and photons they emit during radioactive decay can cause various types of damage to cells, especially to the DNA. Consequently, care must be taken in handling any type of radioactivity.

Natural radioactivity and human mitochondrial DNA mutations. L. Forster *et al.* (2002) *Proc. Natl. Acad. Sci. USA* **99**, 13950–13954.

5.2.A. Alpha Particles

An alpha particle consists of two protons and two neutrons, has a net charge of +2, and is equivalent to a helium nucleus ($^{4}_{2}$He). One alpha particle is usually emitted during decay of the nuclei of heavy radioactive atoms. An example is the radioactive decay of radium-226 to radon-222, which occurs with a half-life of 1600 years:

$$^{226}_{88}\text{Ra} \longrightarrow {}^{222}_{86}\text{Rn} + {}^{4}_{2}\text{He} + 4.78\,\text{MeV} \tag{5.2}$$

During disintegration of radium-226, two electrons are ejected from the outermost electron shell, and the +2 charged helium nucleus is ejected from the nucleus along a straight path. It gives up its energy to its environment and produces ion pairs in the process. Eventually it captures two electrons from its environment and becomes a stable, neutral helium atom. Most of the energy is given up to the absorbing medium as kinetic energy. Ninety-five per cent of the time a 4.70-MeV alpha particle is emitted, and 0.08 MeV is released as recoil energy of radon-222. The remaining 5% of the time a 4.6-MeV alpha particle is emitted, and 0.18 MeV is given off as gamma rays. Emission of an alpha particle decreases the atomic number by 2 and the number of nucleons by 4.

An alpha particle usually travels only a few micrometers in a solid or liquid medium. A layer of skin or a sheet of paper is sufficient to absorb most alpha particles. The greater the atomic number of the atoms in the absorbing material, the greater the absorption of alpha particles.

The tracks of alpha particles may generate secondary electron tracks of low energy, called **delta rays**, which radiate outward from the primary particle track for distances of tens of nanometers.

Radioimmunotherapy with alpha-particle emitting radionuclides. M. R. Zalutsky & O. R. Pozzi (2004) *Q. J. Nucl. Med. Mol. Imaging* **48**, 289–296.

Current status of alpha-particle spectrometry. E. Garciap-Torano (2006) *Appl. Radiat. Isot.* **64**, 1273–1280.

Relative biological effectiveness of the alpha-particle emitter ^{211}At for double-strand break induction in human fibroblasts. A. K. Claesson *et al.* (2007) *Radiat. Res.* **167**, 312–318.

5.2.B. Beta Particles

Beta particles are electrons, ${}^{0}_{-1}e$, that have been ejected from a nucleus. They carry the single negative charge of an electron and have a very small mass that is only about 1/1800th that of a neutron or proton. Emission of a beta particle causes the change of one neutron into a proton in the nucleus. It occurs from radionuclides that have greater numbers of neutrons than protons in their nuclei. An example is the decay of phosphorus-32 to sulfur-32 by emission of a beta particle:

$$ {}^{32}_{15}\mathrm{P} \longrightarrow {}^{32}_{16}\mathrm{S} + \beta + 1.71\,\mathrm{MeV} \tag{5.3} $$

Consequently, the atomic number increases by one but there is no change in the number of nucleons.

Beta particles can be emitted with a continuous distribution of energies, from near-zero to the theoretical maximum. The energy that is not carried by the beta particle is possessed by a neutrino, which is emitted with each beta particle. In a fraction of disintegrations, gamma rays are also emitted and account for some of the energy.

Beta particles can penetrate only a few millimeters in liquid or solid medium. They transfer their energy and negative charge to their environment. The range of beta energies can be measured by adding successively thicker absorbers until the beta particles can no longer be detected.

5.2.C. Positrons

Positrons are positively charged electrons. They are emitted in cases where the atomic nucleus has a low ratio of neutrons to protons and insufficient energy is available for emission of an alpha particle. A positron is emitted when a proton is changed into a neutron within the nucleus, so the atomic number is decreased by one, while the number of nucleons remains unchanged. The emission of positrons is similar to that of beta (minus) particles, which have similar mass and range in tissue and differ only in their charge. A positron quickly combines with an electron from its environment; the two particles annihilate and give off two gamma rays, whose energies of 0.511 MeV are equivalent to the mass of the positron plus the electron.

Quantitating isotopic molecular labels with accelerator mass spectrometry. J. S. Vogel & A. H. Love (2005) *Methods Enzymol.* **402**, 402–422.

5.2.D. Gamma Rays

Gamma rays are photons that are emitted from the nucleus to remove excess energy during radioactive decay. They then have a fixed energy, with a discrete frequency. Gamma rays are usually emitted during beta decay, and they always accompany positron decay. They are highly penetrating electromagnetic photons of energy, like X-rays, but even more energetic, so they are absorbed by matter rather inefficiently. The efficiency of absorption increases with atomic number, but they can still penetrate several centimeters of lead.

An unstable gamma ray-emitting nucleus can decay and give off excess excitation energy by transferring the energy to an orbital K- or L-shell electron, thereby ejecting it from the atom. Outer shell orbital electrons collapse inwards to fill the energy levels vacated by the ejected electrons, and characteristic X-rays are emitted. If these X-rays are absorbed by an inner orbital electron, **internal conversion** may take place and the electron is ejected.

Effects of high-energy electrons and gamma rays directly on protein molecules. E. S. Kempner (2001) *J. Pharm. Sci.* **90**, 1637–1646.

The variation in biological effectiveness of X-rays and gamma rays with energy. M. A. Hill (2004) *Radiat. Prot. Dosimetry* **112**, 471–481.

Test of internal-conversion theory with precise gamma- and X-ray spectroscopy. J. C. Hardy *et al.* (2006) *Appl. Radiat. Isot.* **64**, 1392–1395.

5.3. KINETICS OF RADIOACTIVE DECAY

The rate of radioactive decay is unique to each radioisotope, and each has its own characteristic, constant decay rate. Most importantly, and unlike other chemical reactions, the **rate of decay is independent of the chemical and physical state of the radioisotope and its environment**, including the temperature and the pressure.

Radioactive decay is a unimolecular process and follows first-order kinetics (Section 4.1.A). Consequently, a plot of the logarithm of the number of radioactive disintegrations as a function of time should be linear (Figure 4-2-A).

The time required for any given radioisotope to decay to one-half of its original amount is a measure of the rate at which radioactive transformation takes place. The physical **half-life** ($t_{1/2}$) of a radioactive atom may range from fractions of a second to billions of years and is unique to each radionuclide. Naturally existing radionuclides have long physical half-lives, or are created by the decay of radioisotopes with long half-lives.

The **mean life** of a radioactive material is the time required for it to decay to 1/e (= 0.37) of the original amount. Thus it is the equivalent of the **relaxation time** of first-order chemical reactions (Section 4.1.A.2) and is the inverse of the apparent rate constant.

5.3.A. Units

The original unit of radioactivity was the **curie** (Ci), which was defined as the number of disintegrations per second taking place in 1 g of radium-226:

$$1 \text{ Ci} = 3.7 \times 10^{10} \text{ disintegrations per second} \quad (5.4)$$

This definition was relatively arbitrary, and subsequently the curie has been replaced with the SI unit **bequerel** (Bq), which is one disintegration per second:

$$1 \text{ Ci} = 3.7 \times 10^{10} \text{ Bq} \quad (5.5)$$

The units of curie, millicurie and microcurie are still commonly used. Converting any quantitative measurement of radioactive decay events into curies or bequerels requires the efficiency of the detection system be known. None measure every disintegration.

The **specific activity** is the activity (in units of Bq) of a radioactive material per unit mass or volume, which can refer to the element itself or to the medium in which the radioactive material is contained. The maximum specific activity is that when all the particular atoms are radioactive, which is known as **carrier-free**. This maximum specific activity may be calculated from the rate at which the isotope decays.

5.4. MEASUREMENT OF RADIOACTIVITY

Radioactivity can be measured quantitatively in a variety of instruments that actually count the number of radioactive decay events; alternatively, radioactive molecules can be localized by the consequences of their radioactive decay.

5.4.A. Radiation Counters

Which instruments can be used for counting radioactive decays depends upon the energy emitted.

1. Ionization Monitor

An ionization monitor consists of a small gas chamber containing two electrodes, a voltage supply and a meter. Ionizing radiation that enters through a thin window at one end ionizes the gas, and a pulse of current is recorded on the meter. Such instruments are capable of detecting the decay of ^{32}P and, with special design, ^{14}C, ^{35}S and ^{33}P. More sensitive techniques are required for weak emitters such as ^{3}H (tritium).

A Handbook of Radioactivity Measurements Procedures, 2nd edn (1985) National Council on Radiation Protection and Measurements, MD.

2. Scintillation Counters

Certain substances emit light upon absorption of ionizing radiation, comparable to the fluorescence that can be emitted upon absorption of light. Such **scintillators** (or **fluors**) can be solids, such as sodium iodide, or liquids, such as toluene. Modern scintillation counters use special liquids and scintillators that are available commercially and appropriate for dissolving various types of samples. The light emitted by the scintillator is detected by one or more photomultiplier tubes in the scintillation counter, which have the potential to count each radioactive decay. The measurements can be increased in accuracy by requiring each disintegration to be detected simultaneously by more than one photomultiplier tube. The strength of the signal is related to the energy of the absorbed radiation, making it possible **for the scintillation counter to distinguish between different radioisotopes and to measure them simultaneously in the same sample**.

The technique is complicated because many common substances, even water, can quench the light emission from the scintillators, just like the quenching that occurs in all types of fluorescence. Comparison of measurements therefore requires that all quenching substances in the samples be controlled carefully. Also, light emitted by the Cerenkov effect (Section 5.4.A.3) can complicate the measurements.

Close proximity of the radioactive molecule to the scintillators greatly increases the efficiency of detection, so **scintillation proximity assays** measure binding of a radioactive ligand to an immobilized macromolecule by fixing the macromolecule to a solid matrix containing a scintillator. Binding of the radioactive ligand to the macromolecule produces a substantial increase in the light pulses emitted over when the radioactive ligand is free in solution.

Analysis of triple-label samples by liquid scintillation spectrometry. T. Altzitzoglou (2004) *Appl. Radiat. Isot.* **60**, 487–491.

Standardization of $^{32}P/^{33}P$ and ^{204}Tl by liquid scintillation counting. L. R. Barquero *et al.* (2004) *Appl. Radiat. Isot.* **60**, 615–618.

Measurement of radioligand binding by scintillation proximity assay. J. Berry & M. Price-Jones (2005) *Methods Mol. Biol.* **306**, 121–137.

Application of scintillation proximity assay in drug discovery. S. Wu & B. Liu (2005) *BioDrugs* **19**, 383–392.

Quality control of liquid scintillation counters. F. Jaubert et al. (2006) *Appl. Radiat. Isot.* **64**, 1163–1170.

Synergic quenching effects of water and carbon tetrachloride in liquid scintillation gel samples. A. Grau Carles (2006) *Appl. Radiat. Isot.* **64**, 1505–1509.

3. Cerenkov Radiation

A radioactive source immersed in a pool of water emits blue-white light, which is known as Cerenkov radiation. The importance for molecular biology of the Cerenkov effect is that it is used routinely to measure the amount of the isotope ^{32}P in aqueous samples using a scintillation counter, but without scintillant. The sample is not changed and can subsequently be used in other ways.

Cerenkov radiation results from charged particles traversing a transparent dielectric medium. When the absorbing medium, such as water, interacts with gamma rays from the radioactive source, charged particles are produced and they produce local polarization along their path. The polarized molecules in the medium return to their rest state after passage of the charged particle and emit light if the velocity of the charged particles is greater than that of light in the medium. A wave-front of light is produced from individual molecules and reinforced by constructive interference.

The light pulses emitted can be detected and counted by modern scintillation counters. The light produced in water alone is much less than that produced in the presence of a scintillator, but it can be detected from radionuclides that emit beta particles if their energy is greater than 265 keV. The average energy of beta particles emitted by ^{32}P is 695 keV, so the majority of those emitted can be detected, although the efficiency is only about 30%. Nevertheless, the sample is not altered by counting in this way and it can be used for other purposes, so this method is widely used in molecular biology laboratories to measure the ^{32}P content of samples that are to be used in experiments.

Introduction to Radiological Physics and Radiation Dosimetry. F. H. Attix (1986) John Wiley, NY.

Cerenkov counting. A. BenZikri (2000) *Health Phys.* **79**, S70–71.

Cerenkov counting of low-energy beta-emitters using a new ceramic with high refractive index. M. Takiue *et al.* (2004) *Appl. Radiat. Isot.* **61**, 1335–1337.

5.4.B. Autoradiography

Autoradiography records an image of the beta particles emitted from a preparation containing radioactivity, using films or emulsions that are sensitive to radiation or light. Radioactive samples are placed directly against the film or emulsion to allow beta particles from the sample to create an image in the film emulsion. The highest resolution image is obtained with a radioisotope that emits weak radiation or light that does not travel far.

A photographic film is an embedded suspension of crystals of silver bromide. When the crystals are struck by a charged particle or by light, the silver atoms are ionized to generate a latent image that is invisible. The image is fixed and made visible by using standard photographic developing methods, which remove silver bromide that has not been ionized and leave the remainder as aggregates of reduced silver atoms. These leave a visible dark spot on the film and collectively make up the photographic image.

A single visible silver grain is produced only after it has encountered several ionization events, so **the photographic response would not be directly proportional to the amount of radiation detected**. This nonlinearity of the photographic response can be diminished by **pre-flashing** the film with a uniform low intensity of light; this primes each grain of silver to become reduced and visible after absorbing just one or a very few additional beta particles. The sensitivity of the film is increased enormously, and the photographic response is much more directly proportional to the amount of radiation detected. Exposing the film to radiation at low temperatures can also increase the signal-to-noise ratio.

Autoradiography can provide qualitative information about the images and quantitative information about the amounts of radioactive isotope present, so it has many practical applications in molecular biology. Large, small or microscopic specimens that contain some radiolabeled compound, including sectioned whole organisms, organs, tissues, cells in culture as a monolayer on a glass slide, cellular structures and nucleic acids, can be examined. Coating the sample directly with a radiation-sensitive emulsion, known as **microautoradioagraphy**, provides increased resolution and makes it possible to identify sites within cells that have incorporated radiolabel. Radiolabeled molecules that have been separated by electrophoresis or thin-layer chromatography are frequently detected and quantified by autoradiography. Autoradiography is frequently used to localize proteins on Western blots and hybridized nucleic acids on Southern blots and Northern blots.

An overview on functional receptor autoradiography using [^{35}S]GTPγ. S. J. Sovago *et al.* (2001) *Brain Res. Brain Res. Rev.* **38**, 149–164.

Drug localization and targeting with receptor microscopic autoradiography. W. E. Stumpf (2005) *J. Pharmacol. Toxicol. Methods* **51**, 25–40.

Autoradiography of enzymes, second messenger systems, and ion channels. D. A. Walsh & J. Wharton (2005) *Methods Mol. Biol.* **306**, 139–154.

1. Film-less Autoradiography

X-ray film can be replaced with a variety of radiation detector systems, laser scanners and computer-based imaging systems. Beta-emitting radionuclides are detected by storage phosphor screens that are about 20–100 times more efficient than conventional X-ray film. Consequently, the exposure time is reduced accordingly, and the screens may be processed at room temperature and without a darkroom or the chemicals that are normally required for developing film. Moreover, phosphor screens are reusable.

Both X-ray films and phosphor screens are being replaced by microchannel array detectors, which are about a factor of 10 more efficient than phosphor screens. In addition, they have increased resolution for detecting carbon-14, sulfur-35, phosphorus-32 and iodine-125 from flat specimens.

First images of a digital autoradiography system based on a Medipix2 hybrid silicon pixel detector. G. Mettivier *et al.* (2003) *Phys. Med. Biol.* **48**, N173–181.

Imaging and characterization of radioligands for positron emission tomography using quantitative phosphor imaging autoradiography. P. Johnstrom & A. P. Davenport (2005) *Methods Mol. Biol.* **306**, 203–216.

The physics of computed radiography: measurements of pulse height spectra of photostimulable phosphor screens using prompt luminescence. K. N. Watt *et al.* (2005) *Med. Phys.* **32**, 3589–3598.

2. *Fluorography*

The sensitivity of autoradiography can be enhanced by including a scintillant, known as **fluorography** or **photofluorography**. In this case, the image is produced by light emitted from a fluorescent screen or material when it encounters radiation. The fluorescent screen is coated with fluorescent reagents (fluorophores or luminophores) that cause the screen to emit visible light when struck by ionizing radiation. The fluorescence enhances the intensity of the image recorded in the photographic image and decreases the exposure time. Fluorography is commonly used in blotting experiments. Blotted samples can be detected 3–10 times more readily than using ordinary autoradiography.

Towards a standardized human proteome database: quantitative proteome profiling of living cells. E. Traxler *et al.* (2004) *Proteomics* **4**, 1314–1323.

Detection of co- and posttranslational protein N-myristoylation by metabolic labeling in an insect cell-free protein synthesis system. N. Sakurai *et al.* (2007) *Anal. Biochem.* **362**, 236–244.

5.5. RADIOISOTOPES COMMONLY USED IN MOLECULAR BIOLOGY

In studies of life processes, the most important radioisotopes are those of hydrogen, carbon, sulfur and phosphorus, because these elements are present in practically all cellular components essential to maintaining life. In addition, iodine is very useful because it can be incorporated readily into tyrosine residues of proteins. The more common radionuclides used in biomedical research are described in Table 5-2.

The Radiochemical Manual, 2nd edn (1966) Amersham.

5.5.A. Hydrogen Isotopes

Three isotopes of hydrogen occur naturally: ^{1}H (**protium**), at an abundance of 99.985%, ^{2}H (**deuterium**), at an abundance of 0.015%, plus trace amounts of ^{3}H (**tritium**).

Deuterium isotope effects on noncovalent interactions between molecules. D. Wade (1999) *Chem. Biol. Interact.* **117**, 191–217.

1. *Tritium*

Tritium is the only isotope of hydrogen that is radioactive, with a half-life of 12.35 years. It decays to nonradioactive helium-3 by emission of one beta particle per decay, with a maximum energy of 0.0186 MeV and an average of 0.00568 MeV. Its natural scarcity means that it needs to be synthesized

Table 5-2 The common radioisotopes used in molecular biology: principal radioactive emissions, yields and energies

Element	Mass (amu)[a]	Physical half-life[b]	Beta-particle yield	Average beta energy (MeV)[c]	Gamma ray yield	Gamma energy (MeV)
Tritium	3	12.35 y	1.0	0.00568	–	–
Carbon	11	20.38 m	0.998	0.386	2.0	0.511
Carbon	14	5730 y	1.0	0.0495	–	–
Phosphorus	32	14.29 d	1.0	0.695	–	–
Sulfur	35	87.44 d	1.0	0.076	–	–
Calcium	45	163 d	1.0	0.0771	–	–
Iodine	123	13.22 h	–	–	0.833	0.159
Iodine	124	4.18 d	0.229	0.824	0.453	0.511
					0.611	0.603
					0.101	0.723
					0.105	1.691
Iodine	125	59.6 d	–	–	0.0667	0.0355
Iodine	131	8.021 d	1.0	0.182	0.0606	0.284
					0.812	0.364
					0.0727	0.637

[a] amu, atomic mass units.

[b] y, year; m, month; d, day; h, hour.

[c] MeV, million electron volts.

Data from Knolls Atomic Power Laboratory (1966) *Chart of the Nuclides*, 15th edn, General Electric Co., CA.

in a nuclear reactor, but hundreds of different tritiated compounds are now available commercially and with high specific activity. Molecules are generally labeled with tritium using exchange with $^3H_2$ by heterogeneous or homogeneous catalysis or catalyzed by radiation, or by substitution by chemical reduction or hydrogenation. Tritium may be detected by liquid scintillation counting or by autoradiography.

Tritium is commonly used in molecular biology for tracer studies because of the short range of the weak beta particle it emits, which permits high-resolution autoradiography. Other advantages are its low toxicity and relatively low cost. A serious disadvantage is that the tritium can be lost from a molecule by exchange with solvent water (Section 5.7)and this can be unsuspected. It is widely used in molecular biology as a label for nucleic acid precursors: ^{3}H-thymidine is widely used to monitor DNA replication. It may be injected directly into laboratory animals or added to excised tissues. Tritium-labeled amino acids are frequently used to monitor protein biosynthesis.

Tritium-labeled thymidine and early insights into DNA replication and chromosome structure. J. H. Taylor (1997) *Trends Biochem. Sci.* **22**, 447–450.

Recent developments in tritium incorporation for radiotracer studies. M. Saljoughian & P. G. Williams (2000) *Curr. Pharm. Des.* **6**, 1029–1056.

Facile and efficient postsynthetic tritium labeling method catalyzed by Pd/C in HTO. T. Maegawa *et al.* (2005) *J. Org. Chem.* **70**, 10581–10583.

2. Deuterium

Deuterium is stable and nonradioactive. In molecular biology it is used as heavy water ($^{2}H_2O$ or D_2O) for contrast variation in neutron scattering, altering the density of aqueous solvents, in studying hydrogen exchange reactions in macromolecules, and in simplifying H-NMR spectra, because ^{2}H atoms do not normally give an NMR signal, whereas ^{1}H atoms do.

Biological molecules containing deuterium atoms (^{2}H) in place of hydrogen (^{1}H) are usually produced by growing on deuterated media microorganisms that produce the desired molecules. Relatively high average levels of deuteration (50% for nucleic acids, 70–80% for proteins) can be achieved by growing organisms in $^{2}H_2O$ with normal protiated carbon sources. Typically, however, the aromatic groups are not highly deuterated using this approach. Higher levels of enrichment (>99%) are achieved using $^{2}H_2O$ plus deuterated nutrients. Such nutrients are most readily produced by growing algae on $^{2}H_2O$, then hydrolyzing the cells to their constituent biochemicals. If the production of the protein is sufficiently robust, deuteration can be achieved by growing the producing organism in minimal growth media in $^{2}H_2O$ containing deuterated glucose, glycerol or succinate.

5.5.B. Carbon Isotopes

Thirteen isotopes of carbon are known, ranging in atomic mass number from ^{8}C, with a half-life of only 3×10^{-22} s, to ^{20}C, with a half-life of 0.01 s. Two stable isotopes of carbon occur naturally: ^{12}C, at 98.9% natural abundance, and ^{13}C, at 1.1%. The latter is especially useful in NMR analysis of macromolecules because it gives an NMR signal, whereas the normal isotope, ^{12}C, does not.

Carbon-14 is the most important radioactive isotope of carbon. It has a half-life of 5730 years and decays to stable nitrogen-14 by emitting one beta particle with a maximum energy of 0.156 MeV and an average of 0.0495 MeV. Carbon-14 is best detected by beta-particle liquid scintillation counting, autoradiography or fluorography. Useful quantities of carbon-14 have to be synthesized, usually using neutron reactors. The long half-life of carbon-14 and its method of production result in relatively low specific activities, so high-level carbon-14 labeling of organic materials is difficult. Nevertheless, a great number of ^{14}C-labeled biological reagents are available commercially. Carbon-14 is produced naturally in the atmosphere from nitrogen-14 by cosmic rays, so there is a constant amount of $^{14}CO_2$ in the atmosphere, dictated by its constant rates of formation and radioactive decay. This is the basis for dating biological materials from their content of carbon-14.

Carbon-11 is another radioactive isotope, but with a half-life of only 20.38 min; it decays to boron-11, which is stable, by emitting 0.98 positron particles per decay, with a maximum energy of 0.960 MeV and an average of 0.386 MeV. The emitted positron is annihilated by combining with an electron, which produces two photons with an energy of 0.511 MeV, which can be detected (Section 5.2.C). Carbon-11 is used medically with positron-emission tomography.

Carbon-14 Compounds. J. R. Catch (1961) Butterworths, London.

5.5.C. Phosphorus Isotopes

Only one isotope of element number 15, phosphorus, ^{31}P, occurs naturally, and it is stable. Nevertheless, 17 isotopes are known, ranging from ^{26}P to ^{42}P; most have physical half-lives of seconds or milliseconds and consequently have no practical applications in molecular biology.

The most useful radioactive isotope of phosphorus is phosphorus-32, which has a half-life of 14.29 days. It is usually produced in an accelerator or reactor, with up to 100% yield (carrier-free). It decays to sulfur-32 (Equation 5.3), which is stable, by emitting one beta particle per decay, with a maximum energy of 1.71 MeV and an average of 0.695 MeV. The relatively high energies of its emissions mean that special care must be used in handling this isotope.

Phosphorus-32 is readily detected by liquid scintillation counting, using either scintillants or its Cerenkov radiation, autoradiography and fluorography. Phosphorus-32 is used in molecular biology primarily to label nucleic acids, which have one phosphorous atom per nucleotide, and to monitor the phosphorylation of proteins, which is a very important reversible covalent modification that modulates the biological activities of many proteins.

Another useful radioisotope of phosphorus is phosphorus-33, which has a half-life of 25.4 days. Each decay yields one beta particle, with a maximum energy of 0.1667 MeV and an average of 0.0486 MeV.

Phosphorous-32: practical radiation protection. P. E. Balance *et al.* (1992) H and H Scientific Consultants, Leeds.

5.5.D. Sulfur Isotopes

Thirteen isotopes of sulfur are known, ranging from ^{28}S, with a half-life of 0.12 s, to ^{40}S, with a half-life of 9 s. Four stable isotopes occur naturally: ^{32}S, at 95.02% natural abundance, ^{33}S at 0.75%, ^{34}S at 4.21% and ^{36}S at 0.02%.

The radioactive isotope of sulfur that is most important for molecular biology is sulfur-35. It has a half-life of 87.44 days, decaying to chlorine-35, which is stable, by emitting one beta particle per decay, with a maximum energy of 0.167 MeV and an average of 0.0488 MeV. Sulfur-35 can be detected by liquid scintillation counting, autoradiography and fluorography.

Sulfur-35 can be prepared in an accelerator and in carrier-free form in a reactor. Many ^{35}S-labeled compounds are available commercially. The incorporation of labeled cysteine and methionine is frequently used to monitor protein biosynthesis. High specific activities can be achieved with ^{35}S-labeling, about 200 times greater than that possible with carbon-14. Consequently, sulfur-35 is used in molecular biology whenever possible, and sulfur analogs of the nucleotides have been used frequently in DNA sequencing.

Sulfur. D. D. Dziewiatkowski (1962), in *Mineral Metabolism*, Vol. 2, Part B (C. L. Comar and F. Bronner, eds), Academic Press, NY.

5.5.E. Iodine Isotopes

Iodine is not a natural constituent of many biochemicals but nevertheless it is very useful in molecular biology because it can easily be introduced chemically into proteins, usually in the phenolic ring of tyrosine residues. Iodine is element number 53, and no fewer than 35 isotopes are known; they range from ^{108}I, with a half-life of 36 ms, to ^{142}I, with a half-life of 0.25 s. Only one isotope of iodine (^{127}I) is stable, and it predominates naturally; the next most stable isotope (^{129}I) has a half-life of 15.7 million years. The four most important radioactive isotopes of iodine for use in molecular biology and medicine are ^{123}I, ^{124}I, ^{125}I and ^{131}I (Table 5-2), which are often known collectively as **radioiodine**.

The isotope used most commonly in molecular biology is iodine-125, which can be synthesized in carrier-free form and has a convenient half-life of 60 days. It emits low-energy (0.035 MeV) gamma rays and is used extensively to radiolabel proteins for a variety of applications in molecular biology, including radioimmunoassay, autoradiography and fluorography. Iodine-125 is also commonly used as 5-iodo-2′-deoxyuridine in studies of DNA uptake and hybridization.

Iodine-123 is used as a diagnostic imaging agent in medicine for thyroid imaging and uptake studies. Its relatively short half-life of 13 h requires that it be transported rapidly from its site of production.

Iodine-124 has a half-life of 4.18 days. It is a positron (beta-plus) emitter that has useful applications in positron-emission tomography in diagnostic medicine. It is useful when part of 5-iodo-deoxyuridine, which is an analog of thymidine, and can be incorporated into DNA.

Iodine-131 has a half-life of only 8 days but is used extensively in clinical nuclear medicine and as a label for proteins. It has the disadvantage of emitting many gamma rays and can be a radiation hazard.

All radioiodine compounds must be handled with care, as radioiodine is readily concentrated in the human thyroid gland, where it can cause problems.

Biology of Radioiodine. L. K. Bustad, ed. (1964) Pergamon Press, Oxford.

Improved iodine radiolabels for monoclonal antibody therapy. R. Stein *et al.* (2003) *Cancer Res.* **63**, 111–118.

5.6. KINETIC ISOTOPE EFFECTS

Isotopes differ in their mass but not in their chemical properties. Consequently, they can be useful in kinetic studies of chemical reaction mechanisms, especially in enzyme catalysis. The effect on the rate of the reaction of substituting a different isotope is studied because **lighter isotopes are generally transferred in chemical reactions more rapidly than heavier ones.** This difference is most dramatic with H atoms, for which the reaction rate for transfer of the ^2H and ^3H isotopes may be 1/24 and 1/79, respectively, of that for the ^1H isotope. Smaller isotope effects occur with other atoms, but even in these cases modern sensitive methods of measuring the relative levels of two isotopes, such as mass spectrometry (Chapter 6), can reveal small kinetic differences between two isotopically labeled forms of the reactants.

The extent to which a reaction is slowed by incorporating a heavier isotope demonstrates to what extent that atom is involved in the rate-determining step of the reaction.

The use of isotope effects to determine enzyme mechanisms. W. W. Cleland (2003) *J. Biol. Chem.* **278**, 51974–51984.

The use of isotope effects to determine enzyme mechanisms. W. W. Cleland (2005) *Arch. Biochem. Biophys.* **433**, 2–12.

5.7. ISOTOPE (HYDROGEN) EXCHANGE

The ability of atoms of one isotope of a macromolecule to exchange with another isotope in the solution is a very useful method of characterizing the structure of the macromolecules and its flexibility. The isotopes used most frequently are those of hydrogen, as they can be introduced as labeled forms of water. For example, a protein with H atoms of one isotope is transferred to water of a different isotope, and the exchange between the two is measured. H atoms attached to various other atoms exchange with solvent at different intrinsic rates, depending upon the tendency of the group of ionize. H atoms on O, N or S atoms exchange relatively rapidly, whereas those attached to C atoms do so at much lower rates that are usually not significant.

The exchange reaction between ^1H and ^2H, for example, on a macromolecule can be described by designating the remainder of the macromolecule as M, and O as the O atom of water:

$$^1HM + {}^2HO- \leftrightarrow {}^2HM + {}^1HO- \tag{5.6}$$

This hydrogen exchange reaction has the advantage that the measurement need not perturb the macromolecule and involves only the normal solvent water. Separate samples can be exposed to a labeling pulse for varying time periods, then the exchange quenched to trap the label by adjusting the pH, and the time–course of the loss of label from each site measured. The pulse conditions can be chosen to label selectively the sites of interest for the particular application. The observed rates provide information about the stability and flexibility of the native structure and how it responds to changes in the environment.

Mechanisms and uses of hydrogen exchange. S. W. Englander *et al.* (1996) *Curr. Opinion Struct. Biol.* **6**, 18–23.

Protein analysis by hydrogen exchange mass spectrometry. A. N. Hoofnagle *et al.* (2003) *Ann. Rev. Biophys. Biomolec. Structure* **32**, 1–25.

Native state hydrogen-exchange analysis of protein folding and protein motional domains. C. Woodward *et al.* (2004) *Methods Enzymol.* **380**, 379–400.

5.7.A. Mechanisms of Exchange in Model Molecules

The H atoms of greatest experimental interest are those attached to N atoms, as in the -CO-NH- peptide backbone, the Asn and Gln $-CO-NH_2$ side-chains of peptides and proteins, and the $-NH_2$ and -NH- groups of the nucleic acid bases. When free and accessible to the solvent, these groups exchange in milliseconds or even faster with appropriate catalysts, but much slower, by as much as 10^{12}-fold, when buried in the interior of a macromolecule and inaccessible to solvent. It is these retardations that reveal information about the stability and flexibility of the folded macromolecular structure; the rates at which exchange actually occurs provides information about the molecular dynamics of the molecule. For example, it has provided information about the rates of base-pair opening in the DNA double helix.

The rate of disappearance of H from a macromolecule MH and its replacement by **H** from the solvent is described by:

$$-\frac{d([\mathrm{MH}]-[\mathrm{MH}]_{eq})}{dt} = \frac{d([\mathbf{MH}]-[\mathbf{MH}]_{eq})}{dt} = k_{ex}([\mathrm{MH}]-[\mathrm{MH}]_{eq}) \quad (5.7)$$

where $[\mathrm{MH}]_{eq}$ and $[\mathbf{MH}]_{eq}$ are the equilibrium concentrations of MH and **MH** and k_{ex} is the rate constant for the reaction. Exchange of the amide groups of peptides and proteins is catalyzed by H^+ and OH^- ions, so the value of k_{ex} is described by:

$$k_{ex} = k_H[H^+] + k_{OH}[OH^-] + k_w \quad (5.8)$$

with catalytic rate constants k_H and k_{OH}, respectively. The term k_w accounts for catalysis by water, but this is usually too small to detect, except in solvents less polar than water. The NH group of a typical amide usually has values close to $k_H = 10^{-1}$ $\mathrm{M}^{-1}\mathrm{s}^{-1}$ and $k_{OH} = 10^7$ $\mathrm{M}^{-1}\mathrm{s}^{-1}$. The rate of exchange as a function of pH increases at low and high pH values (Figure 5-1). The minimum rate occurs at the pH specified by:

$$\mathrm{pH}_{min} = {}^1/_2\log(k_H/K_w k_{OH}) \quad (5.9)$$

where

$$K_w = [H^+][OH^-] = 10^{-14}\ \mathrm{M}^2 \quad (5.10)$$

which accounts for the ionization of the water molecule.

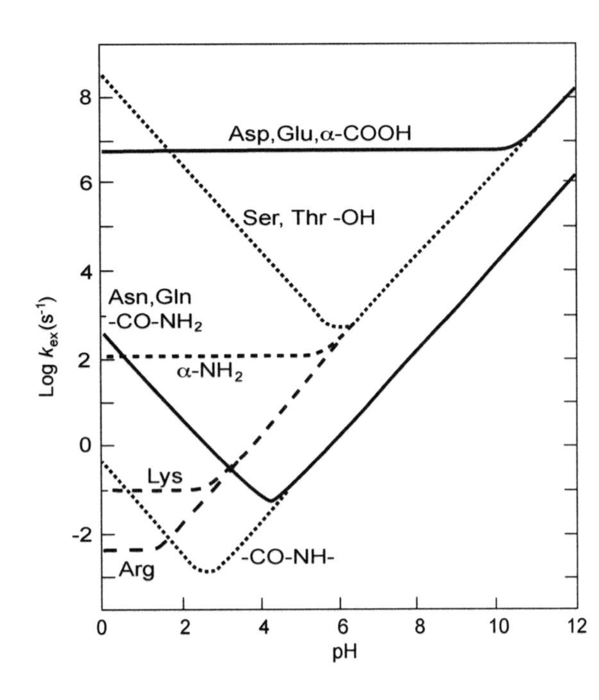

Figure 5-1. Dependence on pH of hydrogen exchange rates (k_{ex}) of model groups in proteins. Data from K. Wüthrich & G. Wagner.

The base-catalyzed exchange of an amide NH occurs by transient removal of the NH H atom, to create the imidate anion (the conjugate base of the amide):

$$\text{R-}\overset{\overset{\displaystyle \text{O}}{\|}}{\text{C}}\text{-NH-R}' + \mathbf{HO}^{-} \rightleftharpoons \underset{\text{imidate}}{\text{R-}\overset{\overset{\displaystyle \text{O}^{-}}{|}}{\text{C}}\text{=N-R}'} + \mathbf{H}\text{OH} \tag{5.11}$$

H atoms from solvent water are depicted bold. The imidate then abstracts an H⁺ from water, to regenerate the amide:

$$\text{R-}\overset{\overset{\displaystyle \text{O}^{-}}{|}}{\text{C}}\text{=N-R}' + \mathbf{H_2}\text{O} \rightleftharpoons \text{R-}\overset{\overset{\displaystyle \text{O}}{\|}}{\text{C}}\text{-N}\mathbf{H}\text{-R}' + \mathbf{H}\text{O}^{-} \tag{5.12}$$

This reaction can occur very rapidly.

The mechanism of the acid-catalyzed reaction depends on the amide. For the $-CO-NH_2$ side-chains of asparagine and glutamine residues of proteins, the exchange occurs simply by transient protonation of the nitrogen, to produce an unstable conjugate acid of the amide, followed by removal of a different hydrogen as H^+:

$$\text{R-}\overset{\overset{\displaystyle \text{O}}{\|}}{\text{C}}\text{-NH}_2 + \mathbf{H}^{+} \rightleftharpoons \text{R-}\overset{\overset{\displaystyle \text{O}}{\|}}{\text{C}}\text{-N}\mathbf{H}\text{H}_2^{+} \rightleftharpoons \text{R-}\overset{\overset{\displaystyle \text{O}}{\|}}{\text{C}}\text{-N}\mathbf{H}\text{H} + \text{H}^{+} \tag{5.13}$$

For backbone –CO–NH– groups, exchange occurs by attachment of H^+ to the O atom, which is much more basic than the nitrogen, to produce another conjugate acid:

$$\text{R-}\overset{\overset{\displaystyle \text{O}}{\|}}{\text{C}}\text{-NH-R}' + \mathbf{H}^{+} \longrightarrow \text{R-}\overset{\overset{\displaystyle \text{O}\mathbf{H}}{|}}{\text{C}}\text{=}\underset{+}{\text{NH}}\text{-R}' \tag{5.14}$$

followed by removal of H^+ from nitrogen, to produce the imidic acid, the unstable tautomer of an amide:

$$\underset{+}{\overset{\mathbf{OH}}{\text{R-C=NH-R}'}} \longrightarrow \underset{\text{imidic acid}}{\overset{\mathbf{OH}}{\text{R-C=N-R}'}} + H^+ \tag{5.15}$$

Attachment of a solvent H^+ to the nitrogen, followed by removal of H^+ from oxygen, regenerates the amide, but with its H exchanged:

$$\overset{\mathbf{OH}}{\text{R-C=N-R}'} + \mathbf{H}^+ \rightleftharpoons \underset{+}{\overset{\mathbf{OH}}{\text{R-C=N}\mathbf{H}\text{-R}'}} \rightleftharpoons \overset{\text{O}}{\overset{\|}{\text{R-C-N}\mathbf{H}\text{-R}'}} + \mathbf{H}^+ \tag{5.16}$$

The rates vary with the substitution pattern in the amide. Table 5-3 lists logarithms of relative rate constants k_H and k_{OH} for various amino acid residues.

Electron-withdrawing groups increase k_{OH} by stabilizing the negative charge that is created in the imidate anion. They also decrease k_H by destabilizing the positive charge that is created in either mechanism. Consequently, the value of pH_{min} (Equation 5.9) is decreased, and the curve in Figure 5-1 moves to the left. The rates of exchange also increase with increasing temperature: in acid, the activation energy is 14 kcal/mol, whereas in base it is 17.5 kcal/mol; this greater value is largely due to the temperature-dependence of the value of K_w.

The intrinsic rate of chemical exchange of amide hydrogens in peptides is influenced by the amino acid sequence surrounding the amide, due to steric and inductive effects of adjacent side-chains that alter the pK_a of the H atom and because of their effects on the local concentration of the proton and hydroxide ion catalysts. These effects have been measured, and using them it is possible to calculate the intrinsic exchange rate expected for any residue in an unfolded polypeptide chain. Because charged species, either cationic or anionic, must be formed as intermediates, the local environment can affect the rate of hydrogen exchange, especially in folded proteins. For example, nearby charges can stabilize or destabilize the charged intermediate by electrostatic interactions. These charges include the phosphate backbone of nucleic acids and the side-chains of acidic or basic residues in proteins, but their effect is moderated by ions of salts that screen the interaction. As a result, the intrinsic rates are affected by the ionic strength of the solution. Moreover, the charges on nearby sites can titrate with pH, leading to a variability of k_{ex} that is more complicated than that illustrated in Figure 5-1. Rates are decreased in less polar environments, where those charged intermediates are less stable.

Similar exchange reactions occur in nucleic acids, except that Equation 5.13 never occurs.

Proton exchange in amides: surprises from simple systems. C. L. Perrin (1989) *Acc. Chem. Res.* **22**, 268–275.

Primary structure effects on peptide group hydrogen-exchange. Y. Bai *et al.* (1993) *Proteins* **17**, 75–86.

Table 5-3. Rate constants for acid- and base-catalyzed hydrogen exchange in the model compound $CH_3C(=O)NH_aCHRC(=O)NH_bCH_3$, relative to R = CH_3

R	$\log_{10}k_{H,a}$	$\log_{10}k_{H,b}$	$\log_{10}k_{OH,a}$	$\log_{10}k_{OH,b}$
CH_3	≡0	≡0	≡0	≡0
H	–0.22	0.22	0.27	0.17
CH_2OH	–0.44	–0.39	0.37	0.30
$CH(OH)CH_3$	–0.79	–0.47	–0.07	0.20
CH_2SH	–0.54	–0.46	0.62	0.55
$(CH_2S-)_2$	–0.74	–0.58	0.55	0.46
$(CH_2)_3NHC(NH_2)_2^+$	–0.59	–0.32	0.08	0.22
CH_2CONH_2	–0.58	–0.13	0.49	0.32
$CH_2CO_2^-$	0.9	0.58	–0.30	–0.18
CH_2COOH	–0.9	–0.12	0.69	0.6
CH_2Imidazole (His)	—	—	–0.10	0.14
CH_2ImidazoleH$^+$	–0.8	–0.51	0.8	0.83
CH_2Phenyl (Phe)	–0.52	–0.43	–0.24	0.06
CH_2Indole (Trp)	–0.40	–0.44	–0.41	–0.11
$CH_2C_6H_4OH$ (Tyr)	–0.41	–0.37	–0.27	0.05
CH_2-CH_2-$CONH_2$	–0.47	–0.27	0.06	0.20
CH_2-$CH(CH_3)_2$	–0.57	–0.13	–0.58	–0.21
CH_2-CH_2-CO_2^-	–0.9	0.31	–0.51	–0.15
CH_2-CH_2-COOH	–0.6	–0.27	0.24	0.39
$(CH_2)_4NH_3^+$	–0.56	–0.29	–0.04	0.12
CH_2-CH_2-SCH_3	–0.64	–0.28	–0.01	0.11
$CH(CH_3)_2$	–0.74	–0.30	–0.70	–0.14
$CH(CH_3)CH_2CH_3$	–0.91	–0.59	–0.73	–0.23
cis-$(CH_2)_3N_\alpha$ (Pro)	—[a]	–0.19	—[a]	–0.24
trans-$(CH_2)_3N_\alpha$ (Pro)	—[a]	–0.85	—[a]	0.60
N-term[b]	—	–1.32	—	1.62
C-term[c]	0.96	—	–1.8	—
C-term[d]	~0.05	—	—	—

[a]H^α absent; [b]for $^+H_3NCH_2CONHCH_3$; [c]for $CH_3C(=O)NHCH_2COOH$; [d]for $CH_3C(=O)NHCH_2CO_2^-$.

Data from Y. Bai *et al.* (1993) *Proteins* **17**, 75–86.

5.7.B. Monitoring Exchange

The rate of hydrogen exchange can be measured most easily by monitoring the overall extent of exchange from the isotopic content of the macromolecule. This might involve counting the radioactivity of tritium or using any absorbance or vibrational spectroscopy that is sensitive to isotope content. These methods measure only the total content of isotope, however, and cannot distinguish between the various H atoms in the macromolecule. Other methods make this possible.

1. NMR

Various types of NMR spectroscopy can distinguish the individual H atoms in a molecule, and each signal can be assigned to a specific H atom. It is also possible to select only particular H atoms, such as those attached to ^{15}N. Only ^{1}H atoms give NMR signals, whereas ^{2}H normally do not. The simplest method for measuring rates of exchange is then to transfer the normal molecule containing ^{1}H to D_2O ($^{2}H_2O$) and to watch the disappearance of each $N^{1}H$ signal (Figure 5-2).

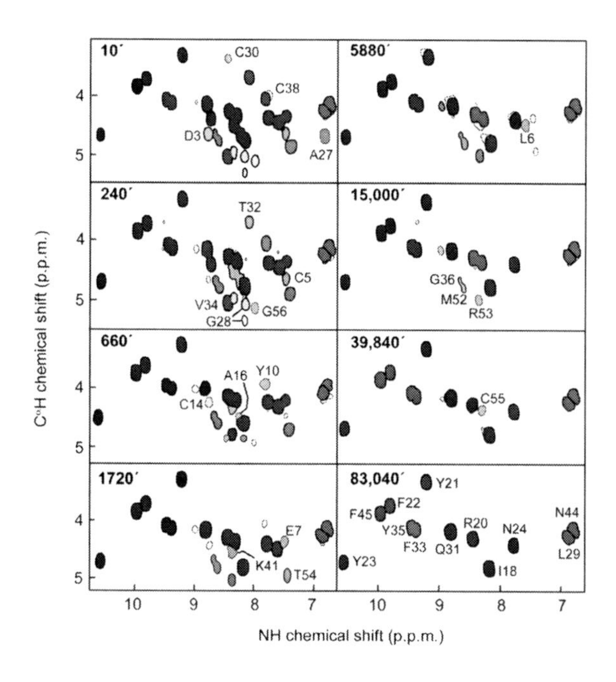

Figure 5-2. Hydrogen exchange of individual backbone amide protons in the protein BPTI followed by the disappearance of the cross-peaks between the NH and C$^{\alpha}$H hydrogens of each residue in two-dimensional COSY NMR spectra. The ^{1}H-labeled protein was dissolved in $^{2}H_2O$ and kept at 36° C for the number of minutes indicated in the top left corner of each spectrum before the spectra were measured. When the $-^{1}$HN– group becomes $-^{2}$HN– as a result of exchange with the solvent, the cross-peak disappears from the spectrum. The magnitude of each cross-peak is indicated here semiquantitatively by the degree of shading. The cross-peaks that disappear completely are identified on the last spectrum in which they are apparent, using the one-letter abbreviation of the amino acid followed by the residue number. The assignments of the most slowly exchanging amides are given on the last spectrum. Data from G. Wagner & K. Wüthrich.

Other techniques use NMR to detect exchange, but they measure rates under equilibrium conditions, when there is no net reaction. In this case, an NH remains at an MH site for an average time of $1/k_{ex}$. The rate of exchange determines the broadening of the NMR signal, which increases with increasing exchange rates until the signal is so broad as to be undetectable. With small molecules, the signal from a –CH group adjacent to an –NH– is split into a doublet. With increasing rate of exchange of the –NH–, the doublet broadens and overlaps, then coalesces into a single peak that is initially broad but then sharpens. Such techniques have the advantage of not being limited by the rapid exchange that can occur during mixing, and can measure very rapid rates of exchange.

2. *Mass Spectrometry*

Mass spectrometry (Chapter 6) can be used to measure the overall exchange process, as the different isotopes alter the molecular weight of the macromolecule slightly: it increases by 1 Da for each deuterium (^2H) that replaces a normal H atom. In Figure 5-3, the molecular weight of the peptide increases with time of incubation of the original protein in ^2H_2O.

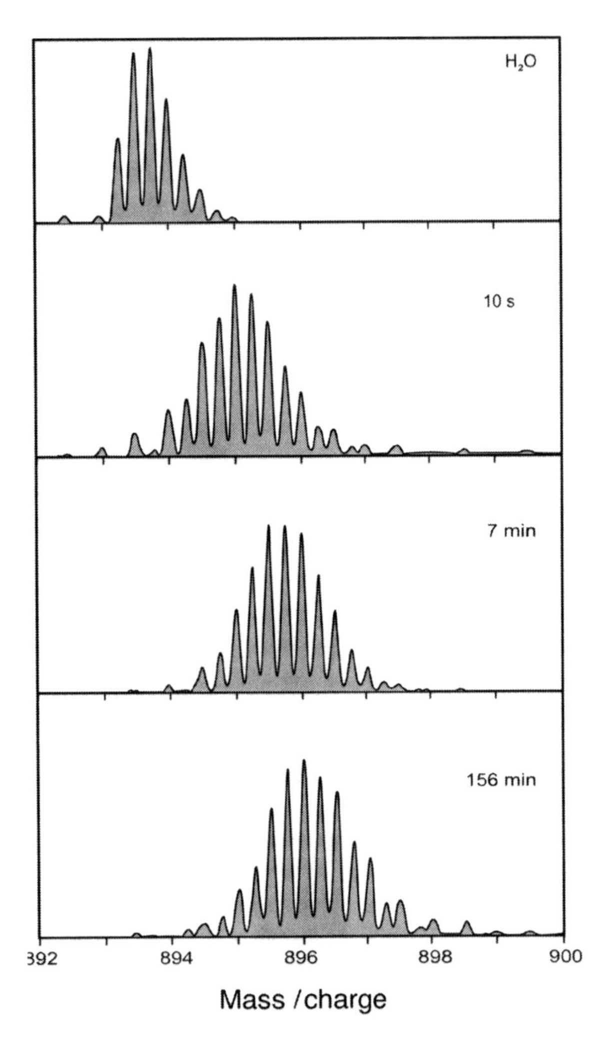

Figure 5-3. Hydrogen exchange monitored by mass spectrometry. The protein was initially present in ^1H_2O and then incubated in ^2H_2O for the indicated periods of time. The protein was digested by pepsin, and one resulting peptide (with net charge of +4 and a normal mass of 3569 Da) analyzed by using a quadrupole orthogonal time-of-flight mass spectrometer and electrospray ionization (Chapter 6). The numerous peaks in each spectrum arise because of the various isotopes present. Data from A. N. Hoofnagle *et al.* (2003) *Ann. Rev. Biophys. Biomol. Structure* **32**, 1–25.

To measure hydrogen exchange rates at individual sites of a macromolecule in this way is difficult, however, because it must be cleaved into small pieces that can be analyzed individually. Conditions that preserve the isotopic label must be maintained, for example using cleavage enzymes that function at 0° C and at pH values where exchange is minimized (Figure 5-1) while the fragments are being separated and analyzed for isotopic content. Some hydrogen exchange during these manipulations is inevitable in practice. The exchanging H atoms can be pinpointed only to the fragment, as the H atoms can migrate on a peptide during the mass spectrum analysis. On the other hand, this approach has the advantages of high sensitivity, wide coverage of the sequence and the ability to analyze large molecules.

Protein analysis by hydrogen exchange mass spectrometry. A. N. Hoofnagle *et al.* (2003) *Ann. Rev. Biophys. Biomol. Struct.* **32**, 1–25.

Methods to study protein dynamics and folding by mass spectrometry. S. J. Eyles & I. A. Kaltashov (2004) *Methods* **34**, 88–99.

Hydrogen exchange and mass spectrometry: a historical perspective. S.W. Englander (2006) *J. Am. Soc. Mass Spectrom.* **17**, 1481–1489.

Hydrogen exchange mass spectrometry for the analysis of protein dynamics. T. E. Wales & J. R. Engen (2006) *Mass Spectrom. Rev.* **25**, 158–170.

3. Neutron Diffraction

Neutrons distinguish readily between ^1H and ^2H atoms, as the former has a negative scattering factor and the latter a large positive one. Consequently, hydrogen exchange can be monitored crystallographically with a macromolecule in a crystal. Crystallographic measurements are not very amenable to rate measurements, so such studies are carried out only when the result is sufficiently important. The results obtained have generally been consistent with those obtained in solution. What is most remarkable is that the rate of exchange is often not perturbed by incorporating the macromolecule into a crystalline lattice.

Hydrogen exchange in RNase A: neutron diffraction study. A. Wlodawer & L. Sjolin (1982) *Proc. Natl. Acad. Sci. USA* **79**, 1418–1422.

Protein dynamics investigated by the neutron diffraction–hydrogen exchange technique. A. A. Kossiakoff (1982) *Nature* **296**, 713–721.

5.7.C. Mechanisms of Exchange in Macromolecules

An NH group at the surface of a macromolecule and hydrogen bonded only to water is usually observed to exchange at a rate like that of an appropriate model small molecule. H atoms that are protected in the interior of a macromolecule and by internal hydrogen bonding exchange more slowly. The degree of this retardation provides information about the stability of the macromolecular structure and the flexibility that permits exchange. In spite of many extensive investigations, however, the detailed mechanism of hydrogen exchange is not known with most macromolecules. Experimental observations are usually discussed in terms of two limiting and extreme models: (1) **solvent penetration** of the structure and (2) **local unfolding** (or breathing) of the structure. For exchange to occur, water molecules of the solvent must come into contact with the H atom to be exchanged. In the first case, the water molecule goes to the exchangeable site on or in the molecule, whereas in the second the molecule unfolds sufficiently to become accessible to the solvent.

1. Solvent Penetration Model

According to this model, water and catalytic H^+ or OH^- gain access to the interior of the macromolecule through channels, perhaps generated transiently by conformational fluctuations. The rate of exchange is then governed by the extent to which those buried H atoms come into contact with the solvent atoms and by the environment within the altered protein structure; the nature of the environment where exchange takes place is unknown, so rates of exchange cannot be interpreted quantitatively, and this model is not considered very useful. It does, however, account more readily for experimental observations with proteins.

2. Local Unfolding Model

According to the **local unfolding** model, flexibility of a segment of the molecule breaks a set of hydrogen bonds and other interactions that hold the macromolecule in its stable conformation. The NH groups of that segment are thereby exposed to solvent, in a transient 'open' conformation, where they can undergo exchange with the bulk solvent.

The local unfolding model can be analyzed kinetically:

$$\text{closed} \underset{k_{fold}}{\overset{k_{unfold}}{\rightleftharpoons}} \text{open} \xrightarrow{k_{ex}} \text{exchanged product} \tag{5.17}$$

Here k_{unfold} is the rate constant for the motion that converts the closed form of the macromolecule to the open one, k_{fold} is the rate constant for reversion back to the closed form, and k_{ex} is the intrinsic rate constant for hydrogen exchange (Equation 5.8). The general expression for the observed rate constant is:

$$k_{obs} = \frac{k_{unfold}\, k_{ex}}{k_{fold} + k_{ex}} \tag{5.18}$$

Two limiting mechanisms can be distinguished, depending upon the relative magnitudes of k_{ex} and k_{fold}.

a. EX1 Mechanism

If $k_{ex} >> k_{fold}$, Equation 5.18 simplifies to:

$$k_{obs} = k_{unfold} \tag{5.19}$$

The rate-limiting step is the opening of the protein structure, to produce the locally unfolded state, from which exchange is rapid. This is the EX1 mechanism. Under these conditions, the rate of exchange should be independent of the pH, except insofar as k_{unfold} is pH-dependent. These conditions can be induced at high pH, where k_{ex} becomes sufficiently great. In practice, however, it is rare to achieve these conditions for a protein without denaturation becoming significant; then exchange generally occurs from the totally unfolded conformation. Nucleic acids are much more stable at alkaline pH and can be studied under EX1 conditions, which made make it possible to measure the frequencies with which base pairs are broken transiently.

b. EX2 Mechanism

If $k_{fold} >> k_{ex}$, Equation 5.18 simplifies to:

$$k_{obs} = K_{op} k_{ex} \tag{5.20}$$

where K_{op} is the equilibrium constant for forming the locally unfolded state:

$$K_{op} = k_{unfold}/k_{fold} = [open]/[closed] \tag{5.21}$$

This is the EX2 mechanism. Exchange occurs only during that fraction of the time when the macromolecule is unfolded. The value of k_{ex} will be approximately the same as the observed value for an appropriate small molecule model under the same conditions; the experimental value of k_{obs} provides an estimate of K_{op}. This value can then be converted to the free energy change involved in the local unfolding:

$$\Delta G° = -RT \log_e K_{op} \tag{5.22}$$

This value is an estimate of the energy that normally stabilizes the closed native state of the macromolecule, or of the energy required to allow the local unfolding. Alternatively, the degree to which the exchange is retarded by the protein structure can be expressed as a **protection factor**, P:

$$P = k_{ex}/k_{obs} = 1/K_{op} \tag{5.23}$$

Protein hydrogen exchange mechanism: local fluctuations. H. Maity *et al.* (2003) *Protein Sci.* **12**, 153–160.

EX1 hydrogen exchange and protein folding. D. M. Ferrarro *et al.* (2004) *Biochemistry* **43**, 587–594.

~ CHAPTER 6 ~

MASS SPECTROMETRY

A major concern in molecular biology is determining the size of a macromolecule. Proteins and nucleic acids have a very wide range of sizes, and it is important to know whether one is working with a large molecule or a relatively small one. Consequently, a great many techniques have been devised to provide some estimate of the sizes of macromolecules, including light scattering, microscopy, diffusion and viscosity, sedimentation, polyacrylamide gel electrophoresis, and size exclusion chromatography. The most accurate technique by far, however, is mass spectrometry (MS).

Mass spectrometry requires that the sample be in the gas phase and ionized, by either the loss or the gain of at least one charge, due to electron ejection, protonation or deprotonation. The mass spectrometer then separates the ions on the basis of their ratio of mass (m) to charge (z) (Figure 6-1). The molecules being separated must not collide with other molecules or atoms, so the separation must take place in a high vacuum. The result is a mass spectrum that can provide very accurate measures of the molecule's mass. Even structural information can be inferred from the charge distribution on a protein. In some cases, the ionization process breaks the molecule into pieces, which can yield information about its covalent structure.

The ions are detected by either **electron multipliers** or **photomultipliers**. The ion strikes either a dynode that emits secondary electrons or a phosphorous screen that releases photons. The secondary electrons of the dynode are accelerated by a voltage and attracted to a second dynode that emits more electrons. Passage through additional dynodes can amplify the original signal about 10^6-fold. The photons released by a phosphorous screen are detected by a photomultiplier.

Molecules are distinguished in mass spectrometry solely on the basis of their masses, which are expressed in Daltons (Da, 1 Da = 1 g/mol).*

---

* *The absolute masses of molecules are measured in Daltons (Da). Mass spectrometers, however, measure relative masses, because the instrument has been calibrated with molecules of known mass. Such masses are relative and should have no units, but they should have the same magnitude as the absolute mass, so they are commonly referred to as absolute masses. The term 'molecular weight' is widely used, even though strictly speaking a weight is a force (mass × gravity), not a mass, but it has the same magnitude as the mass and is used very commonly to refer to the mass.*

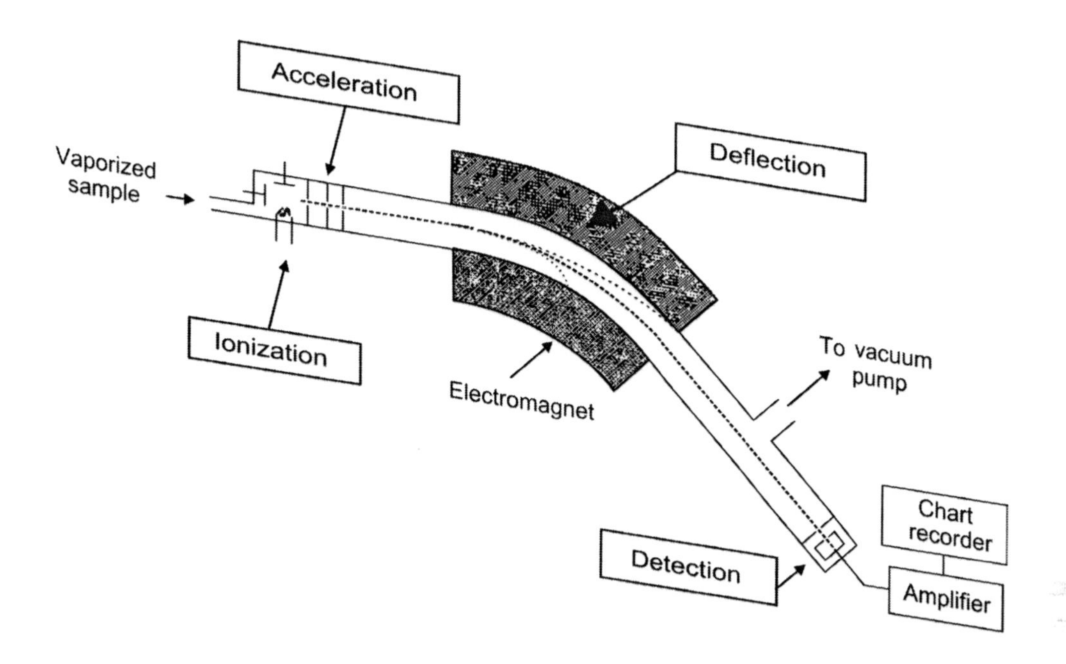

Figure 6-1. Schematic diagram of a mass spectrometer. The molecules of the sample are ionized by knocking off one or more electrons to give a positive ion. The ions are accelerated in an electric field to have the same kinetic energy, then deflected by a magnetic field. The degree of deflection depends upon the ratio of their mass to their net charge: lighter ions are deflected more than heavier ones with the same net charge. Which ions reach the detector depends upon the strength of the magnetic field, which is varied to detect a range of molecules. When positive ions reach the detector, they acquire electrons from it, which is detected by an electric current to the detector. So that the flights of the molecules are not affected by collisions with gas molecules, all of this must take place in a vacuum.

Masses of the natural amino acid residues and nucleotides are unique, except for the amino acid residues Leu and Ile, which are isomers and have the same molecular weight. They can be distinguished by mass spectrometry only if they differ in the way that the polypeptide chain is fragmented.

Mass spectrometry has been used routinely for many years with small molecules, but its use with biological macromolecules was prevented for a long time by the need to generate intact ions in the gas phase; biological macromolecules are not detectably volatile. The developments of **electrospray ionization (ESI)** and **matrix-assisted laser desorption/ionization (MALDI)** overcame this problem, and mass spectrometry has become an integral and important part of biological research. ESI and MALDI are fundamentally different ionization techniques, but they achieve essentially the same end result, namely the generation of gas-phase ions of the molecules of the sample by their vaporization and ionization without destroying them. Ionization occurs in both techniques by the addition or abstraction of protons from the molecule M, to produce either $[M+H]^+$ or $[M-H]^-$ ions. Peptides and proteins are generally studied as $[M+H]^+$ ions, nucleic acids as $[M-H]^-$ ions.

Proteins, peptides, carbohydrates and oligonucleotides can now be analyzed routinely by mass spectrometry and examined structurally in very small (picomole to femtomole) amounts. Mass spectrometry can be used to measure the masses of very large biomolecules, up to 10^6 Da in the case of proteins, but about 90 kDa for DNA and 150 kDa for RNA. It can also provide sequence information on unknown peptides and proteins and detect noncovalent complexes, with a molecular weight accuracy of the order of ±0·01% or better. With very high accuracy measurements, the various minor isotopes present naturally, such as ^{13}C and ^{15}N (Section 5.1), become apparent (Figure 5-3). Proteins and nucleic acids with masses greater than about 8000 Da have at least one such isotope in virtually every molecule.

Molecular weight determination of peptides and proteins by ESI and MALDI. K. Strupat (2005) *Methods Enzymol.* **405**, 1–36.

Mass spectrometry of peptides and proteins. V. H. Wysocki *et al.* (2005) *Methods* **35**, 211–222.

Mass spectrometry and protein analysis. B. Domon & R. Aebersold (2006) *Science* **312**, 212–217.

Mass spectrometry of RNA. B. Thomas & A. V. Akoulitchev (2006) *Trends Biochem. Sci.* **31**, 173–181.

Mass spectrometry of RNA: linking the genome to the proteome. Z. Meng & P. A. Limbach (2006) *Brief Funct. Genomic Proteomic* **5**, 87–95.

6.1. ELECTROSPRAY IONIZATION (ESI)

ESI generates ions of macromolecules directly from an aqueous or aqueous/organic solvent that does not contain much salt. A fine spray of highly charged droplets is created in the presence of a strong electric field (Figure 6-2). These charged droplets vaporize as they move into the vacuum of the mass spectrometer, which concentrates the ionized molecules. When the electrostatic repulsion between them becomes sufficiently great, they leave the droplet and enter the gas phase individually as charged ions. The number of charges on a molecule depends on factors such as the composition and pH of the solvent and the chemical nature of the sample. Proteins with multiple positive charges are usually obtained by ESI from acidic solutions of pH 2–4, while negatively charged protein molecules are produced from alkaline solutions of pH 8–10. ESI usually produces a series of multiply charged species of large molecules. Proteins from acidic solution often contain one positive charge on each accessible basic group, i.e. those of the lysine and arginine side-chains, plus the terminal amino group.

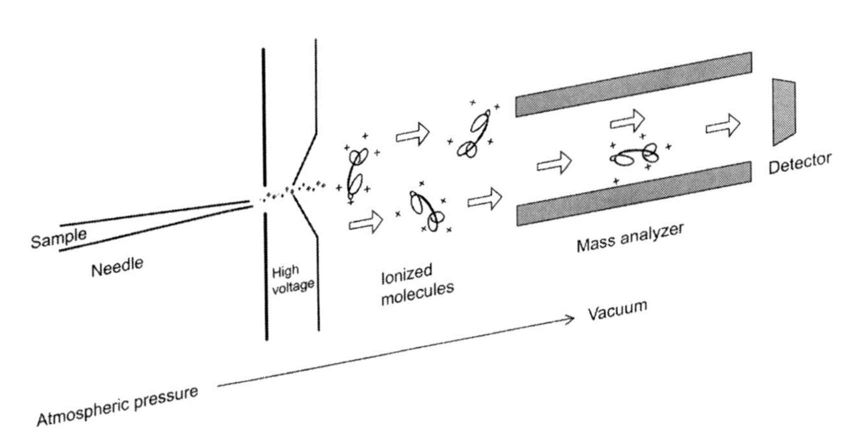

Figure 6-2. The electrospray ionization process. An aqueous solution of the analyte (at a concentration of about 5 μM) is placed in the needle. A voltage of several kilovolts is applied to the gold-plated needle, producing an electrospray of fine droplets from 1 to 2 μl of the sample. The positively charged droplets are desolvated by the vacuum, and the individual molecules are separated and detected by the mass spectrometer.

The **ESI mass spectrum of a homogeneous macromolecule contains multiple peaks corresponding to the different charged states** and different mass-to-charge (*m*/*z*) ratios (Figure 6-3). Adjacent peaks differ by one integral charge, plus one proton, so the spectrum can be deconvoluted to determine the net charge of each peak and the molecular mass of the original molecule.

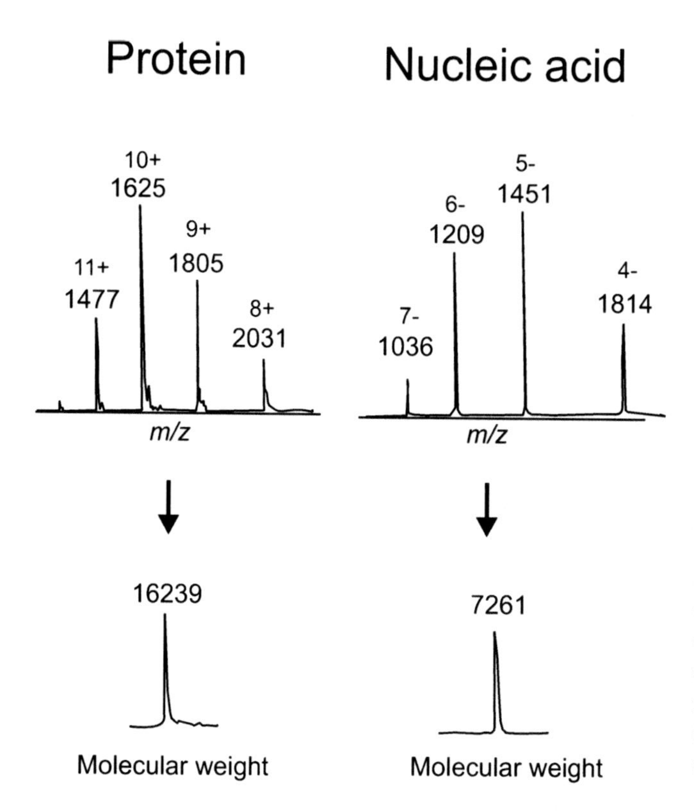

Figure 6-3. Examples of data generated with an ESI mass spectrometer. Proteins usually produce ions with multiple positive charges (*left*), while oligonucleotides generate ions with multiple negative charges (*right*). The number of charges and the *m*/*z*-value measured for each peak are given. Below each spectrum is an example of the type of molecular weight information generated by deconvoluting the data above. The molecular weight of the original molecule is given.

The multiple charging that occurs in ESI is a unique and useful characteristic of the technique. It enables a molecule's mass to be determined with great precision, because masses can be calculated independently from several different charged states. The multiple charging also permits the analysis of large molecules, even using conventional mass analyzers that are normally limited to the detection of relatively small molecules. For example, a 70-kDa protein subjected to ESI will typically contain 40–70 charges. This will produce multiply charged species with *m*/*z* ratios of between only 1000 and 2000, which can be detected readily with quadrupole mass analyzers (Section 6.3.B).

ESI is the gentlest ionization technique, and **the ionization of a macromolecule by ESI can reflect its conformational properties**. For example, folded proteins usually have smaller numbers of charges than unfolded proteins, and the two forms can be distinguished (Figure 6-4). Complexes in which one molecule binds to another noncovalently can survive the ionization, persist in the gas phase, and be detected. In some cases, however, complexes of the macromolecule with salt ions are detected.

Another advantage of ESI-MS is that it is compatible with liquid chromatography, so molecules emerging from a chromatography column can be injected directly into the mass spectrometer.

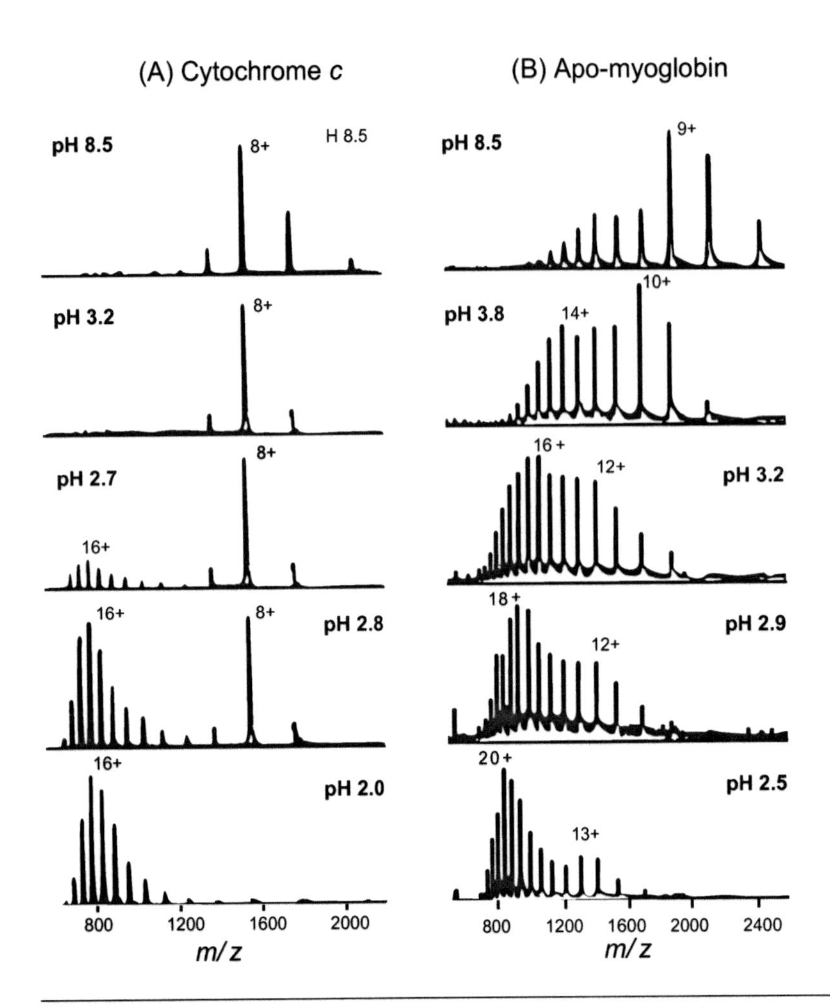

Figure 6-4. ESI spectra of cytochrome *c* (*left*) and apo-myoglobin (*right*) when initially folded and unfolded. The samples were initially at the indicated pH values. At pH 8.5, both proteins are folded in their native conformations. At acidic pH values, however, both unfold. In both cases, the unfolded form has a much greater net charge than the folded form. Native cytochrome *c* molecules generally have a net charge of +8, whereas the unfolded molecules average +16. In the region of pH 2.7, only two populations of molecules appear to be present, either folded or unfolded. Apo-myoglobin, in contrast, appears to have an intermediate form populated from pH 3.8 to 2.5, as the charge spectrum of the molecules gradually shifts. Data from L. Konermann & D. J. Douglas.

Recent developments in electrospray ionisation mass spectrometry: noncovalently bound protein complexes. A. E. Ashcroft (2005) *Nat. Prod. Rep.* **22**, 452–464.

Future directions for electrospray ionization for biological analysis using mass spectrometry. R. D. Smith (2006) *Biotechniques* **41**, 147–148.

Protein structures under electrospray conditions. A. Patriksson *et al.* (2007) *Biochemistry* **46**, 933–945.

6.2. MATRIX-ASSISTED LASER DESORPTION/IONIZATION (MALDI)

Gas-phase ions are generated using MALDI by the vaporization of a mixture of the molecule of interest in a solid matrix upon radiation with intense light from a laser (Figure 6-5). The macromolecule is mixed and embedded in the solid matrix, which often consists of an organic material that absorbs light, such as *trans*-3-indoleacrylic acid, and inorganic salts, such as sodium chloride and silver trifluoroacetate. This solid sample is then irradiated with a laser producing light with a wavelength that is absorbed by the matrix. Short laser pulses of 10–20 ns duration and a power of about 10^6 W/cm^2 eject electronically excited matrix ions, cations and neutral macromolecules into the gaseous phase. The macromolecules become ionized by collisions with small cations, such as H^+, Na^+ and Ag^+. They usually acquire relatively few charges.

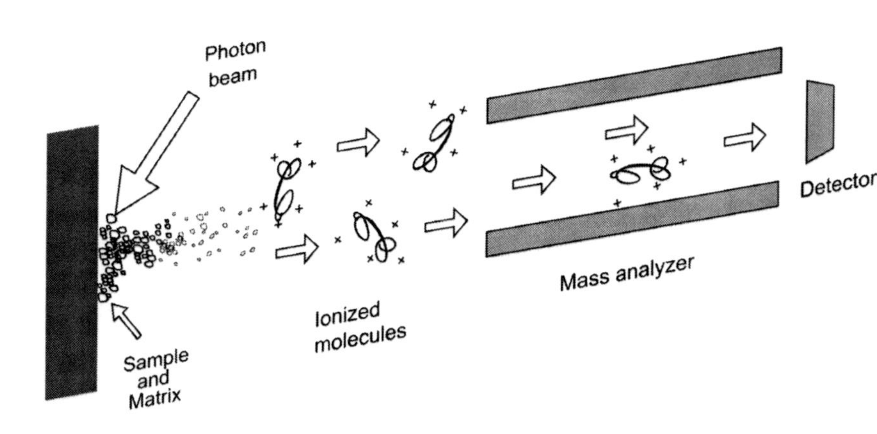

Figure 6-5. The MALDI process. The photon beam is absorbed by the matrix and ejects ionized macromolecules that were embedded in it; they are analyzed in the mass analyzer.

Low molecular-weight molecules of less than 20 kDa typically acquire only one or two charges, while larger molecules can have as many as 3–5, depending on the specific conditions used, the type of matrix material and the power of the laser. The relatively low number of charge states observed in MALDI makes the technique especially well-suited for the analysis of multi-component mixtures, because individual components can be identified easily by the signal generated by their +1 charge state (Figure 6-6). Dimers and trimers of molecules, plus complexes with materials of the matrix, can also be observed in the spectrum. Molecules with masses of up to 300,000 Da can be analyzed.

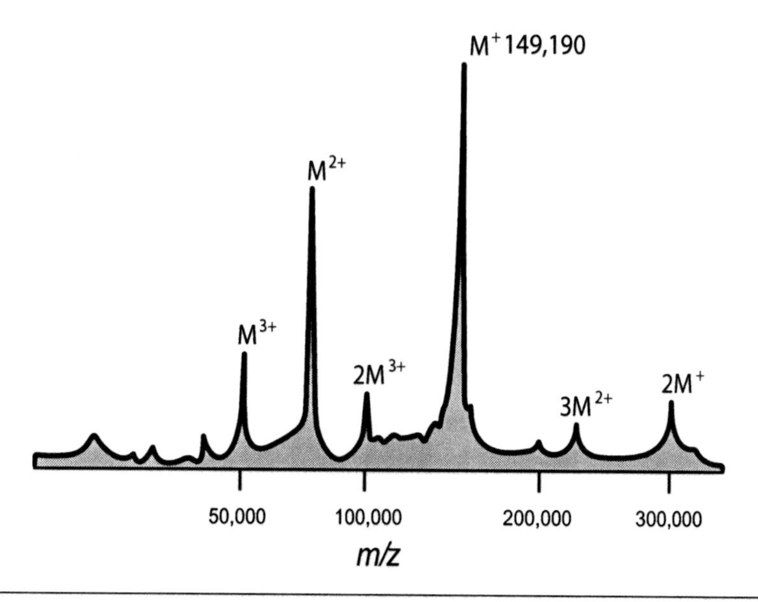

Figure 6-6. The MALDI mass spectrum of a monoclonal antibody. Monomers (M), dimers (2M) and trimers (3M) are apparent, with up to 3+ charges. Data from F. Hillenkamp & M. Karas.

MALDI: more than peptide mass fingerprints. K. Stuhler & H. E. Meyer (2004) *Curr. Opinion Mol. Ther.* **6**, 239–248.

DNA analysis by MALDI-TOF mass spectrometry. I. G. Gut (2004) *Human Mutat.* **23**, 437–441.

MALDI-TOF mass spectrometry: a versatile tool for high-performance DNA analysis. C. Jurinke *et al.* (2004) *Mol. Biotechnol.* **26**, 147–164.

Matrix-assisted laser desorption/ionisation, time-of-flight mass spectrometry in genomics research. J. Ragoussis *et al.* (2006) *PLoS Genet.* **2**, e100.

6.3. MASS ANALYZERS

Mass analyzers separate ions according to their mass to charge ratio (m/z); they are the most crucial part of a mass spectrometer.

ESI-MS commonly uses **quadrupole** mass analyzers, whereas MALDI-MS uses **time-of-flight** (TOF) mass analyzers. The resolution offered by TOF mass analyzers is less than that of quadrupole mass analyzers. This, combined with the complication of adduct formation, results in MALDI-MS having a lower accuracy, of the order of 0.1%, whereas ESI-MS typically has an accuracy of roughly 0.01%. Higher resolution mass analyzers, such as the ultra-high-resolution **Fourier-transform ion cyclotron resonance mass analyzer**, to produce FTMS (Section 6.3.D), give accuracies of better than 0.001%.

6.3.A. Magnetic Focusing

The mass spectrometer shown in Figure 6-1 separates ions with varying ratios of m/z by their different trajectories within a magnetic field. The radius r followed by ions with a particular m/z depends upon the accelerating voltage (V) and the strength of the magnetic field (H) according to:

$$r = \frac{1}{\mathrm{H}}\left(2\mathrm{V}\frac{m}{z}\right)^{1/2} \tag{6.1}$$

Consequently, varying the strength of the magnetic field permits ions with varying m/z-values to pass through the detector and generate the mass spectrum. Good accuracy requires that all the ions have the same energy after acceleration, which is accomplished in **double-focusing magnetic sector spectrometers** by subjecting the ions to a constant voltage either prior to or subsequent to the magnetic field. This produces high resolution and the ability to analyze molecules with m/z-values of up to 15,000 Da, but these instruments are very expensive.

6.3.B. Quadrupole Mass Analyzers

A quadrupole mass analyzer is depicted in Figure 6-7. Only electric fields, not magnetic, are used to separate the ions. The quadrupole consists of four parallel metal rods, and the ions pass down the middle between them. The four rods are linked in two pairs, A and B, and each pair is connected in order to have the same electrical properties. The two pairs have opposite constant direct-current (dc) voltages. Superimposed on each dc voltage is an oscillating voltage, with opposite phases in the A and B rods. For any particular electrical field, only ions with a specified value of m/z pass between the rods and through the slit to the detector. The other ions are deflected and collide with the rods. All the ions in the sample are detected sequentially by varying the applied voltages, permitting ions with varying values of m/z up to 4000 to be detected.

A quadrupole mass spectrometer can be coupled with an **ion trap**, in which the ions are held within another type of quadrupole and manipulated before being permitted to reach the detector. The ions are physically trapped between the electrodes and subjected to both constant and oscillating electric fields, so that ions of specific m/z precess within the trap. As the magnitudes of the electric fields are varied, ions of various m/z are ejected from the trap and allowed to reach the detector. The ion trap increases both the resolution and the sensitivity, and molecules with masses of up to 100,000 Da can be analyzed with an accuracy of up to 0.003%.

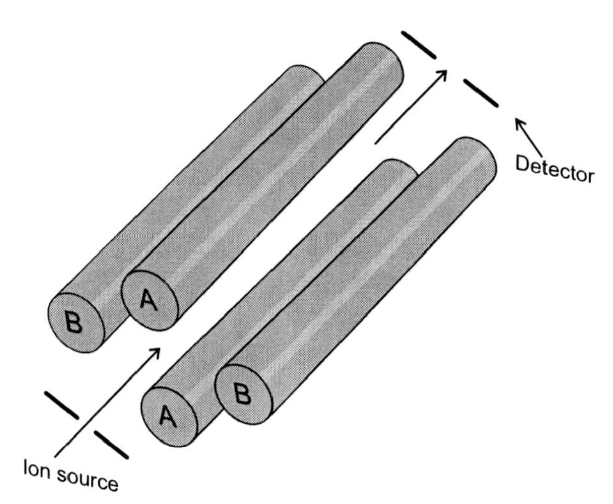

Figure 6-7. A quadrupole mass analyzer. The ions pass between the four parallel rods, which have a fixed voltage, plus one varying with radio frequency. Rods labeled A are connected and have the same voltages, as are those labeled B. The B rods have the opposite fixed voltage to the A rods, and the radio frequency phase is shifted 180° from the A rods. Depending upon the voltages, most ions are deflected; only those with a certain ratio of *m*/*z* continue through the middle of the rods and pass through the slit to the detector.

Quadrupole and quadrupole ion trap mass analyzers are ideally suited to be coupled with ESI; they have three primary advantages. First, they are tolerant of relatively high pressures of up to 10^{-7} atmospheres, which is well-suited to electrospray ionization, where the ions are produced under atmospheric pressure conditions. Secondly, quadrupoles and ion traps can analyze molecules with an *m*/*z* ratio of up to 4000, within the range produced by ESI of proteins and other biomolecules. Finally, these mass analyzers are relatively inexpensive, so most successful commercial electrospray instruments use quadrupole mass analyzers.

C-terminal peptide sequencing using acetylated peptides with MS^n in a quadrupole ion trap. A. H. Payne *et al.* (2000) *Analyst* **125**, 635–640.

Tandem mass spectrometry in quadrupole ion trap and ion cyclotron resonance mass spectrometers. A. H. Payne & G. L. Glish (2005) *Methods Enzymol.* **402**, 109–148.

6.3.C. Time-of-Flight (TOF) Analyzers

One of the simplest mass analyzers is the TOF analyzer. A set of ions is accelerated towards a detector with the same amount of energy. If the ions have the same energy, but different masses, they will have different velocities and reach the detector at different times. The smaller ions will reach the detector first because of their greater velocities; the larger ions will take longer. Their mass is determined by the ions' flight time through the analyzer. TOF analyzers are commonly used with MALDI ionization.

The ions are accelerated over a short distance *d* by an electrical field of strength E and then travel through a drift region of length *l*. The time, *t*, required for an ion of mass *m* and charge number *z* to reach the detector at the end of the drift region is given by:

$$t = l\,(m/2z\,\mathrm{eE}d)^{1/2} \tag{6.2}$$

where e is the unit of fundamental electrical charge. Because all the other parameters are fixed and known, the ratio *m*/*z* is determined by the time of flight, *t*. The difference in the time of arrival of ions is not great, usually in the microsecond time range, so a complete spectrum can be measured in a very short time. For example, an accelerating voltage of 20 kV will cause a singly charged ion with a mass of 1000 Da to have a velocity of about 6×10^4 m s^{-1}, and the time to travel 1 m will be 1.7×10^{-5} s.

The ions must enter the flight tube at exactly the same time, which is generally accomplished by generating ions in short bursts, using a pulsed laser with MALDI. The sensitivity and resolution can be increased by slowing the ions with a series of electric field 'lenses' until they stop, then accelerating them in the opposite direction. This 'reflection' increases the path length the ions travel, thereby increasing their separation. The lenses also focus the ions with a specific m/z by reducing the spread in their kinetic energies.

TOF analyzers have the advantages of being very sensitive and able to analyze molecules of essentially any mass.

Tandem time-of-flight mass spectrometry. M. L. Vestal & J.M. Campbell (2005) *Methods Enzymol.* **402**, 79–108.

6.3.D. Fourier-Transform Ion Cyclotron Resonance (FTMS)

FTMS offers high resolution and the ability to perform experiments involving multiple collisions, MS^n, where n can be as high as 4 (Section 6.4), but it requires a cyclotron and a super-conducting magnet. The ions are injected into a small volume in the cyclotron, and a strong magnetic field is applied so that the ions precess in circular orbits that depend upon the magnetic field and their m/z ratio. The ions are kept within the cell by an electric field that is applied to front and rear plates of the sample cell. They are not detected directly, but by their absorption of energy when subjected to an electric field with a frequency that matches their precession frequency, analogous to what happens in magnetic resonance experiments. The ions transmit a radio frequency current at the detector plates that contains the frequency components of each of the ions. This is converted to a free ion decay signal, which can be transformed into the mass spectrum.

FTMS has the ability to analyze very large molecules, with masses of > 10^6 Da. Combined with ESI and MALDI, FTMS offers high accuracy, with errors less than ±0.001%.

Examples of Fourier transform ion cyclotron resonance mass spectrometry developments: from ion physics to remote access biochemical mass spectrometry. A. Rompp *et al.* (2005) *Eur. J. Mass Spectrom.* **11**, 443–456.

Nucleic acid analysis by Fourier transform ion cyclotron resonance mass spectrometry at the beginning of the twenty-first century. J. L. Frahm & D. C. Muddiman (2005) *Curr. Pharm. Des.* **11**, 2593–2613.

Protein primary structure using orthogonal fragmentation techniques in Fourier transform mass spectrometry. R. Zubarev (2006) *Expert Rev. Proteomics* **3**, 251–261.

6.4. TANDEM MASS SPECTROMETRY (MS^n)

The development of tandem mass spectrometry combined with ESI has made determining the complete sequences of biopolymers such as proteins and oligonucleotides routine. ESI generates the intact molecular ion in the gas phase, which is then collided with neutral atoms such as argon or helium.

Collision-induced dissociation (CID) results, producing fragment ions that can be analyzed by their masses. This approach of inducing fragmentation and performing successive mass spectrometry experiments on the fragment ions is known as **tandem mass spectrometry**. It is usually abbreviated as MS^n, where n refers to the number of generations of fragment ions being analyzed (Figure 6-8). The sequence of the peptide can be assembled from the masses of the fragments produced because the collision-induced fragmentation of peptides is well-characterized. Tandem mass spectrometry is used routinely to acquire partial or total sequences of small peptides with fewer than 30 amino acid residues, and of short oligonucleotides.

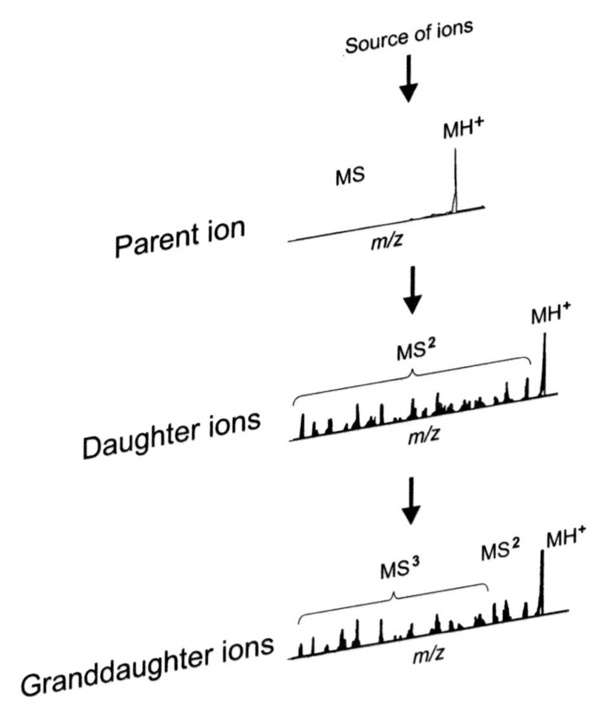

Figure 6-8. Tandem mass spectrometry: generation of fragment ions via collision-induced dissociation and mass analysis (MS^n) of the progeny fragment ions. The parent ion is selected on the basis of its mass. For the MS^2 experiment, the molecular ion MH^+ can be selected by the analyzer and caused to undergo collision-induced dissociation that results in its fragmentation; the products are then analyzed. In an MS^3 experiment, a daughter fragment ion is selected and exposed to collision-induced dissociation, generating granddaughter fragment ions. The terms 'parent', 'daughter' and 'granddaughter' ions are used here, but 'precursor', 'product' and 'second-generation product' ions are also commonly used terms.

Electron capture dissociation (ECD) is a new fragmentation technique that is used in Fourier transform ion cyclotron resonance mass spectrometry (Section 6.3.D) that is complementary to traditional tandem mass spectrometry techniques. Fragmentation is fast and specific, and labile post-translational modifications and noncovalent bonds often remain intact after backbone bond breakage. Disulfide bonds are normally stable to vibrational excitation but are cleaved preferentially in ECD. ECD provides extensive sequence information with polypeptides, and at high electron energies even Ile and Leu residues are distinguishable.

Detection and localization of protein modifications by high resolution tandem mass spectrometry. F. Meng *et al.* (2005) *Mass Spectrom. Rev.* **24**, 126-134.

Tandem mass spectrometry for peptide and protein sequence analysis. J. J. Coon *et al.* (2005) *Biotechniques* **38**, 519, 521, 523.

Analysis of posttranslational modifications of proteins by tandem mass spectrometry. M. R. Larsen *et al.* (2006) *Biotechniques* **40**, 790–798.

INDEX

Page numbers for major entries, illustrations and tables are in **bold**. Page numbers for items that are primarily a reference are in *italics*.